Collins

Maths
Frameworking

3rd edition

Kevin Evans, Keith Gordon,
Trevor Senior, Brian Speed,
Chris Pearce

Collins

Contents

How to use this book

Learning objectives

See what you are going to cover and what you should already know at the start of each chapter.

About this chapter

Find out the history of the maths you are going to learn and how it is used in real-life contexts.

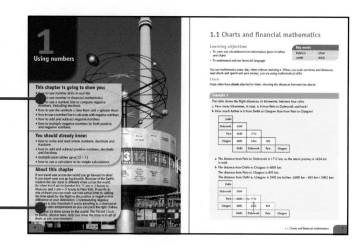

Key words

The main terms used are listed at the start of each topic and highlighted in the text the first time they come up, helping you to master the terminology you need to express yourself fluently about maths. Definitions are provided in the glossary at the back of the book.

Worked examples

Understand the topic before you start the exercises, by reading the examples in blue boxes. These take you through how to answer a question step by step.

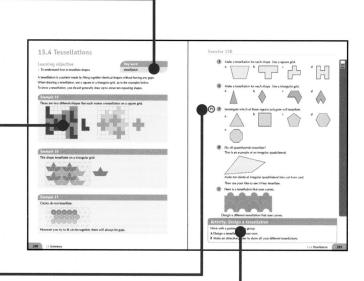

Skills focus

Practise your problem-solving, mathematical reasoning and financial skills.

Take it further

Stretch your thinking by working through the **Investigation**, **Problem solving**, **Challenge** and **Activity** sections. By tackling these you are working at a higher level.

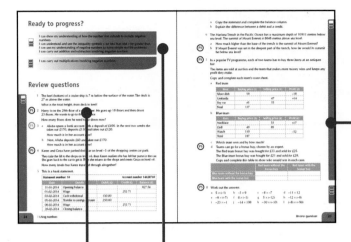

Progress indicators

Track your progress with indicators that show the difficulty level of each question.

Ready to progress?

Check whether you have achieved the expected level of progress in each chapter. The statements show you what you need to know and how you can improve.

Review questions

The review questions bring together what you've learnt in this and earlier chapters, helping you to develop your mathematical fluency.

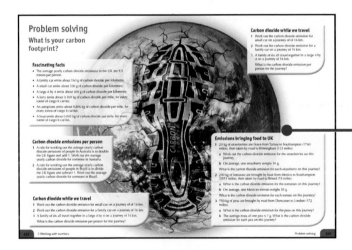

Activity pages

Put maths into context with these colourful pages showing real-world situations involving maths. You are practising your problem-solving, reasoning and financial skills.

Interactive book, digital resources and videos

A digital version of this Pupil Book is available, with interactive classroom and homework activities, assessments, worked examples and tools that have been specially developed to help you improve your maths skills. Also included are engaging video clips that explain essential concepts, and exciting real-life videos and images that bring to life the awe and wonder of maths.

Find out more at www.collins.co.uk/connect

1
Using numbers

This chapter is going to show you:

- how to use number skills in real life
- how to use number in financial mathematics
- how to use a number line to compare negative numbers, including decimals
- how to use the symbols < (less than) and > (greater than)
- how to use a number line to calculate with negative numbers
- how to add and subtract negative numbers
- how to multiply negative numbers by both positive and negative numbers.

You should already know:

- how to write and read whole numbers, decimals and fractions
- how to add and subtract positive numbers, decimals and fractions
- multiplication tables up to 12 × 12
- how to use a calculator to do simple calculations.

About this chapter

If you travel east across the world you go forward in time! If you travel west you go backwards. Because of the Earth's rotation the day starts at different times across the world. So when it is 8 am in London it is 11 am (+ 3 hours) in Moscow and 3 am (– 5 hours) in New York. If you fly to one of them you can work out your arrival time by adding the time taken by the flight to the positive or negative time difference at your destination. Understanding negative numbers is also important if you're travelling to a destination with sub-zero temperatures so you can pack the right clothes.

There are 24 time zones in the world. The World Clock in Berlin, shown here, tells you what the time is in all of them at any one moment!

1.1 Charts and financial mathematics

Learning objectives

- To carry out calculations from information given in tables and charts
- To understand and use financial language

Key words

balance	chart
credit	debit

You use mathematics every day, often without realising it. When you work out times and distances, read charts and spend and save money, you are using mathematical skills.

Charts

Maps often have **charts** attached to them, showing the distances between key places.

Example 1

The table shows the flight distances, in kilometres, between four cities.

a How many kilometres, in total, is it from Paris to Dubrovnik and back?

b How much further is it from Delhi to Glasgow than from Paris to Glasgow?

Delhi				
Dubrovnik	5099			
Paris	6580	1712		
Glasgow	6885	2426	893	
	Delhi	Dubrovnik	Paris	Glasgow

a The distance from Paris to Dubrovnik is 1712 km, so the return journey is 3424 km in total.

b The distance from Delhi to Glasgow is 6885 km.

The distance from Paris to Glasgow is 893 km.

The distance from Delhi to Glasgow is 5992 km further. (6885 km − 893 km = 5992 km)

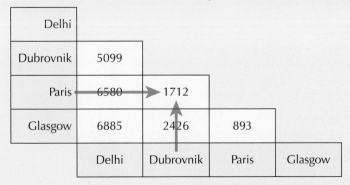

Planning and making purchases

Information about prices is often given in tables. This makes it easier to read.

Example 2

A park hires out bicycles. These are the prices.

Hire period	Bicycle	Tandem
2 hours	£7.50 adult	£13.50
	£5.00 child	
4 hours	£12.50 adult	£19.50
	£7.50 child	
All day	£15.00 adult	£25.00
	£10.00 child	

a How much does it cost to hire bicycles for 4 hours for 1 adult and 2 children?

b How much more does it cost to hire two tandems for a whole day than for 2 hours?

   a Total cost for 4 hours for 1 adult and 2 children is £12.50 + £7.50 + £7.50 = £27.50.

   b One tandem costs £25.00 for a whole day but only £13.50 for 2 hours.

      The difference is £25.00 − £13.50 = £11.50.

      The cost for two tandems will be £11.50 × 2 = £23 more.

Bank statements

A bank statement gives you detailed information about your bank account. It shows details of money that has been paid in or out of the account, and the amount of money remaining.

A **debit** is the amount paid out of an account.

A **credit** is the amount paid in to an account.

The **balance** is the amount of money remaining in the account.

Example 3

Here is a bank statement.

Statement number: 9   **Account number 13579246**

Date	Details	Debit (£)	Credit (£)	Balance (£)
31-01-2014	Opening balance			417.83
01-02-2014	Interest		15.41	
03-02-2014	Cash withdrawal	180.00		
05-02-2014	The music shop	9.79		
26-02-2014	Salary		354.68	
28-02-2014	Closing balance			598.13

a Copy the bank statement and complete the balance column.

b How much was paid out from the account in February?

a

	Statement number: 9			Account number 13579246	
Date	Details	Debit (£)	Credit (£)	Balance (£)
31-01-2014	Opening balance			417.83
01-02-2014	Interest		15.41	433.24
03-02-2014	Cash withdrawal	180.00		253.24
05-02-2014	The music shop	9.79		243.45
26-02-2014	Salary		354.68	598.13
28-02-2014	Closing balance			598.13

b £189.79 was paid out.

Example 4

Here are two readings from a gas meter.

January 2014

April 2014

The readings give the number of metric units of gas that have been used.

Charges are based on kilowatt hours (kWh).

To convert metric units to kilowatt hours, multiply the number of units by 11.2.

The first 670 kWh are charged at 8.40p per kWh.

The remainder are charged at 5.00p per kWh.

a How many metric units of gas were used in the period from January to April?

b Work out the cost of the gas used.

  a The number of units used is 24 569 – 24 401 = 168 metric units.

  b 168 metric units = 168 × 11.2 kWh

                = 1881.6 kWh

  670 kWh at 8.40p = £56.28

  1881.6 – 670 = 1211.6 kWh

  1211.6 at 5.00p = £60.58

  The cost of the gas used is £56.28 + £60.58 = £116.86.

Exercise 1A

(FS) (1) A TV and broadband package costs £23.50 a month for the first 6 months and then £49.99 per month.

Work out the total cost for the first two years.

(FS) (2) Work out the total cost of this mobile phone contract over the length of the plan.

Include the cost of the phone in your total.

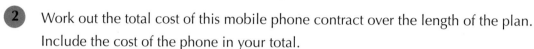

- Unlimited internet and texts
- 2000 minutes to any network
- Plus unlimited calls
  Just £36.00 a month
  - Phone cost £75
- Plan length 18 months

(FS) (3) **a** Copy this bank statement.

Then fill in the balance column.

Date	Details	Debit (£)	Credit (£)	Balance (£)
31-01-2014	Opening balance			326.25
01-02-2014	Interest		8.21	
05-02-2014	Shirt shop	53.62		
05-02-2014	The hungry cafe	16.88		
05-02-2104	Birthday shop	22.79		
26-02-2014	Paid in		228.54	
28-02-2014	Closing balance			

b How much was paid out on the 5 February?

c What is the difference between the opening and closing balances?

(FS) (4) Four friends agreed to deposit a fixed amount each month into their bank accounts.

Copy and complete the table.

Name	Heather	Iain	Joanna	Kenny
Opening bank balance	£222.22	£194.63	£133.95	£96.80
Amount saved per month	£17.50	£22.50	£30.00	£48.00
Amount saved in 12 months				
Closing bank balance				

Who had the most money at the end of 12 months?

5 This chart shows the distances between four cities in England.

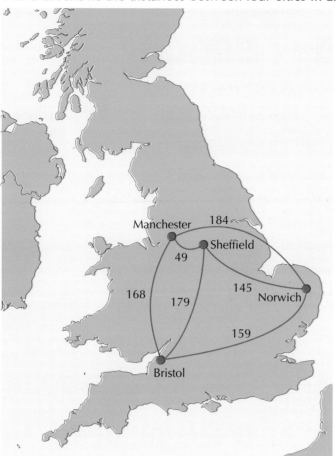

a Copy and complete the mileage chart.

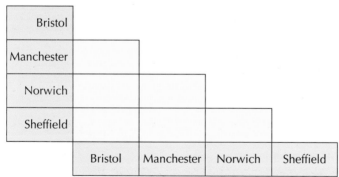

Bristol			
Manchester			
Norwich			
Sheffield			
Bristol	Manchester	Norwich	Sheffield

  6 Work out the shortest route to visit all four cities. Start from Sheffield.

  7 These are two readings from an electricity meter. The units are given in kWh.

January 2014

April 2014

The first 150 kWh are charged at 20.8p per kWh.

The remainder are charged at 12.5p per kWh.

a How many kWh of electricity were used between January and April?

b Work out the cost of the electricity used.

 **8** These are two readings from a gas meter.

January 2014

April 2014

The readings give the number of metric units used.

The first 670 kWh are charged at 8.40p per kWh.

The remainder is charged at 5.00p per kWh.

> **Hint** To convert metric units to kilowatt hours, multiply the number of units by 11.2.

a How many metric units of gas were used in the period from January to April?

b Work out the cost of the gas used.

Activity: Sending a parcel

Use an internet search to find the cheapest way to send a 15 kg parcel from the United Kingdom to Poland. Assume the parcel is 30 cm long, 15 cm wide and 20 cm high.

1.2 Positive and negative numbers

Learning objectives

- To use a number line to order positive and negative numbers, including decimals

- To understand and use the symbols < (less than) and > (greater than)

Look at the two pictures.

What are the differences between them?

Every number has a sign. Numbers greater than 0 are called **positive numbers**. Although you do not always write it, every positive number has a positive (+) sign in front of it.

Numbers less than 0 are called **negative numbers** and must always have a negative (−) sign in front of them.

The positions of positive and negative numbers can be shown on a number line. The value of the numbers increases as you move from left to right. For example, −5 is **greater than** −10, 2 is greater than −5 and 8 is greater than 2.

Key words

greater than	less than
negative number	positive number

Temperature +26.5 °C

Temperature −13.0 °C

−10 −9 −8 −7 −6 −5 −4 −3 −2 −1 0 1 2 3 4 5 6 7 8 9 10

You can use the number line to compare the sizes of positive and negative numbers. You can also use it to solve problems involving addition and subtraction.

Example 5

Which number is greater, −7 or −3?

Because −3 is further to the right on the number line than −7 is, on the number line, it is the larger number.

Notice that −3 is closer to zero than −7 is.

Example 6

Write these temperatures in order from lowest to highest.

8.5 °C, −2.4 °C, 10.1 °C, −7.0 °C, −3.5 °C, 4.8 °C

Putting these temperatures on a number line you can see the correct order.

−7.0 °C, −3.5 °C, −2.4 °C, 4.8 °C, 8.5 °C, 10.1 °C

The symbol > means greater than. The symbol < means **less than**.

To remember which symbol is less than, notice that < looks similar to the letter L.

For example:

−4 < 7 means 'negative 4 is less than 7'.

−3 > −8 means 'negative 3 is greater than negative 8'.

Example 7

State whether each statement is true or false.

a 6.5 > 8.1   **b** −7.2 > −1.8   **c** −3.4 > −3.8

Putting each of these pairs on the number line shows that:

a 6.5 > 8.1 is false

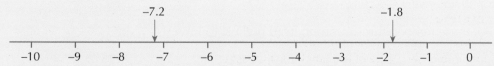

b −7.2 > −1.8 is false

c −3.4 > −3.8 is true.

Exercise 1B

1 Write down the highest and lowest temperatures in each group.

a   −4 °C, −2 °C, 0 °C   b   −8 °C, −15 °C, −10 °C   c   −20 °C, −19 °C, −15 °C

2 Work out the difference between the temperatures in each pair.

a   −8 °C and 17 °C   b   −13 °C and −25 °C   c   14 °C and −7 °C

3 On Monday the temperature at noon was 2 °C.
Over the next few days these temperature changes were recorded.

Monday to Tuesday	down eight degrees
Tuesday to Wednesday	up three degrees
Wednesday to Thursday	down five degrees
Thursday to Friday	up nine degrees

What was the temperature on Friday?

4 Put these numbers in order, from smallest to largest.

a   13, −8, 2, −7, 9   b   −11, −7, 8, −12, −10   c   0, −4, −6, −11, 4

d   9, −13, 8, −9, −14   e   −7, −9, −18, 10, −10   f   19, −8, 7, −17, 5

5 State whether each statement is true or false.

a $7.5 > 3.8$ b $2.9 < 16.1$ c $5.8 < −6.2$ d $−8.6 > −5$

e $−2.7 < −9.1$ f $−7.2 > 1.3$ g $−4.3 < −3.5$ h $−9.3 < 3$

6 Copy each statement and put < or > into the ☐ to make it true.

a   −5.3 ☐ 4.2   b   −7.8 ☐ −10.6   c   3.2 ☐ −3.5   d   −12.6 ☐ −2.4

(PS) 7 Work out the number that is halfway between the numbers in each pair.

a   ⊢───┬───┬───⊣
       −17   2

b   ⊢───┬───┬───⊣
       −9   7

c   ⊢───┬───┬───⊣
       −23   −7

8 Put these temperatures in order, from highest to lowest.

15.5 °C, −4.6 °C, 15.8 °C, −4.9°C, −3.5 °C

9 Work out the differences between the temperatures in each pair.

a   −4 °C and 6 °C   b   −2 °C and −4 °C   c   7 °C and −8 °C

10 On one day in December the temperature was 2 °C in London and −4 °C in Edinburgh.

How much lower was the temperature in Edinburgh than in London?

11 a Copy this bank statement.

Then fill in the balance column.

Date	Details	Debit (£)	Credit (£)	Balance (£)
31-01-2014	Opening balance			187.00
05-02-2014	Cash withdrawal	53.62		
18-02-2014	Cash withdrawal	228.54		
28-02-2014	Closing balance			

b Explain why the closing balance is negative.

Challenge: Changing state

A The table shows the temperatures at which some substances change from being solids to become liquids.

Fluid	Temperature (°C)
Butane	−138
Carbon dioxide	−79
Castor oil	−10
Chloroform	−64
Ether	−116
Glycerine	−8
Linseed oil	−20
Mercury	−39
Nitrogen	−210
Turpentine	−59
Water, fresh	0
Water, sea	−3

Copy the table and list the temperatures, in order of size.

Make sure that the highest temperature is at the top and the lowest is at the bottom.

B Which substances are liquid at −60 °C?

C Solid nitrogen melts at −210 °C.
How many degrees warmer does it need to be before solid carbon dioxide melts?

1.3 Simple arithmetic with negative numbers

Learning objectives

<table>
<tr><td colspan="2">Key words</td></tr>
<tr><td>add</td><td>brackets</td></tr>
</table>

- To carry out additions and subtractions involving negative numbers

- To use a number line to calculate with negative numbers

You can use a number line to **add** and subtract positive and negative numbers.

Example 8

Use a number line to work out the answers.

a 5 – 13   **b** (–11) + 9   **c** 6 – 12 – 3

  a Starting at zero and 'jumping' along the number line to 5 and then back 13 gives an answer of –8.

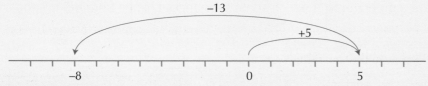

  b Similarly, (–11) + 9 = –2

Notice that **brackets** are sometimes used so that the negative sign is not confused with a subtraction sign.

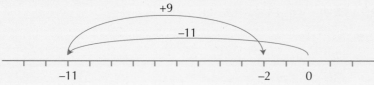

  c Using two steps this time, 6 –12 – 3 = –9

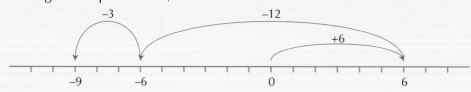

Look at these patterns.

8 + 2 = 10	2 + 8 = 10
8 + 0 = 8	0 + 8 = 8
8 + (–2) = 6	(–2) + 8 = 6
8 + (–4) = 4	(–4) + 8 = 4

Notice that 8 + (−2) = 6 and (−2) + 8 = 6 have the same value as 8 − 2 = 6

and 8 + (−4) = 4 and (−4) + 8 = 4 have the same value as 8 − 4 = 4.

Adding a negative number gives the same result as subtracting the corresponding positive number.

Example 9

Work out the answers.

a 5 + (−3) **b** 20 + (−4) **c** (−5) + (−2)

a 5 + −3 = 5 − 3 **b** 20 + −4 = 20 − 4 **c** (−5) + (−2) = (−5) − 2

$\quad\quad\quad$ = 2 $\quad\quad\quad\quad\quad\quad\quad\quad$ = 16 $\quad\quad\quad\quad\quad\quad\quad\quad$ = −7

Exercise 1C

1 Work out the answers.

a 8 − 19 $\quad$ **b** 4 − 13 $\quad$ **c** 12 − 15 $\quad$ **d** 13 + 19

e 21 − 13 $\quad$ **f** 34 − 34 $\quad$ **g** −11 + 21 $\quad$ **h** −9 − 12

i −17 − 13 $\quad$ **j** −16 + 8 $\quad$ **k** −12 − 14 $\quad$ **l** −18 + 6

m −32 − 23 + 24 $\quad$ **n** −17 + 21 − 32 $\quad$ **o** −23 + 14 − 27 $\quad$ **p** −102 + 103 − 95

2 Copy and complete these calculations.

a 16 + (−7) $\quad$ **b** 28 + (−13) $\quad$ **c** 26 + (−17) $\quad$ **d** 26 + (−15) $\quad$ **e** (−26) + (−27)

$\quad$ = 16 − 7 $\quad\quad$ = 28 − 13 $\quad\quad$ = 26 − ☐ $\quad\quad$ = ☐☐☐ $\quad\quad$ = ☐☐☐

$\quad$ = ☐ $\quad\quad\quad$ = ☐ $\quad\quad\quad\quad$ = ☐ $\quad\quad\quad\quad$ = ☐ $\quad\quad\quad\quad$ = ☐

3 Use the number line below to work out the answers.

a 13 − 15 $\quad$ **b** 18 + (−12) $\quad$ **c** 4 + (−15) $\quad$ **d** 13 + (−3)

e (−12) + (−3) $\quad$ **f** 12 − 20 $\quad$ **g** (−14) + 20 $\quad$ **h** 0 − 15

i 12 + (−15) $\quad$ **j** (−8) + (−6) $\quad$ **k** 12 + (−12) $\quad$ **l** 15 + (−15)

m 14 + (−20) $\quad$ **n** 15 + (−25) $\quad$ **o** 0 + (−11) $\quad$ **p** (−11) + (−4)

−15 −14 −13 −12 −11 −10 −9 −8 −7 −6 −5 −4 −3 −2 −1 0 1 2 3 4 5 6 7 8 9 10 11 12 13 14 15

4 Work out the answers.

a 24 + (−15) $\quad$ **b** (−25) + (−30) $\quad$ **c** 70 − 98 $\quad$ **d** (−17) + (−28)

e 53 + (−17) $\quad$ **f** (−60) + 60 $\quad$ **g** 45 + (−60) $\quad$ **h** 124 − 242

i 113 + (−98) $\quad$ **j** (−140) + (−25) $\quad$ **k** 36 + (−55) $\quad$ **l** (−19) + (−29)

5 Work out the total of the numbers in each list.

a 15, −24, 17, −8, −19, 23 $\quad$ **b** −12, 20, 35, −38, −45, 20

6 In each magic square, all the rows, columns and diagonals add up to the same total. Copy and complete the squares.

a

16		6
	0	
		−16

b

		−24
	−15	−9
−6		

c

0		
−20		
−16		−24

 7 Alf has £124 in the bank. He writes a cheque for £135.

How much has he got in the bank now?

Problem solving: Magic squares

A In this 4 × 4 magic square, all of the rows, columns and diagonals add to −18.
Copy and complete the square.

−27			15
6			−12
		18	−3
	3	−15	

B In this 4 × 4 magic square, all of the rows, columns and diagonals add to the same number.
Copy and complete the square.

0		−26	−6
	−10		
−14	−18	−20	
−24			−30

1.4 Subtracting negative numbers

Learning objective

• To carry out subtractions involving negative numbers

Key word

subtract

Look at this pattern of numbers.

Notice that $8 - (-2) = 10$ has the same value as
$8 + 2 = 10$

and $8 - (-3) = 11$ has the same value as
$8 + 3 = 11$.

Subtracting a negative number is the same as adding a positive number.

$$8 - 3 = 5$$
$$8 - 2 = 6$$
$$8 - 1 = 7$$
$$8 - 0 = 8$$
$$8 - (-1) = 9$$
$$8 - (-2) = 10$$
$$8 - (-3) = 11$$

Example 10

Work out the answers.

a $12 - (-15)$ **b** $23 - (-17)$

a $12 - (-15) = 12 + 15 = 27$ **b** $23 - (-17) = 23 + 17 = 40$

Exercise 1D

1 Copy and complete these calculations.

a $16 - (-12)$ **b** $13 - (-16)$ **c** $22 - (-18)$ **d** $-24 - (-15)$ **e** $(-16) - (-27)$
  $= 16 + 12$ $= 13 + 16$ $= 22 + \square$ $= \square$ $= \square$
  $= \square$ $= \square$ $= \square$ $= \square$ $= \square$

2 Use the number line below to work out the answers.

a $13 - (-2)$ **b** $7 - (-8)$ **c** $(-14) - (-9)$ **d** $6 - (-9)$
e $(-12) - (-8)$ **f** $(-12) - (-10)$ **g** $(-14) - (-20)$ **h** $(-11) - (-19)$
i $8 - (-5)$ **j** $(-13) - (-6)$ **k** $(-15) - (-8)$ **l** $(-12) - (-6)$
m $(-15) - (-7)$ **n** $(-14) - (-25)$ **o** $0 - (-11)$ **p** $(-13) - (-7)$

$-15\ -14\ -13\ -12\ -11\ -10\ -9\ -8\ -7\ -6\ -5\ -4\ -3\ -2\ -1\ \ 0\ \ 1\ \ 2\ \ 3\ \ 4\ \ 5\ \ 6\ \ 7\ \ 8\ \ 9\ \ 10\ 11\ 12\ 13\ 14\ 15$

3 Work out the answers.

a $30 - (-18)$ **b** $(-25) - (-25)$ **c** $32 - (-100)$ **d** $(-17) - (-17)$
e $29 - (-18)$ **f** $36 - (-36)$ **g** $-21 - (-43)$ **h** $-350 - (-290)$
i $106 - (-78)$ **j** $(-123) - (-78)$ **k** $36 - (-45)$ **l** $(-18) - (-49)$

4 Work out the answers.

a $27 + 16 - (-38)$ **b** $(-42) - 31 - (-18)$
c $340 - (-123) + (-91)$ **d** $(-102) + 31 - (-50)$

(PS) **5** Choose a number from each list and subtract one from the other. Repeat for at least four pairs of numbers. What are the biggest and smallest answers you can find?

A	14	−17	−25	11	15
B	−23	9	−18	8	−14

6 Copy and complete these calculations.

a $17 + \square = 6$ **b** $30 - \square = 12$ **c** $\square + (-15) = 18$ **d** $(-28) - \square = 14$

7 Copy and complete these calculations, following the pattern each time.

a $9 + +5 = 14$ **b** $-7 - +2 = -9$ **c** $16 - +8 = 8$
$9 + 0 = 9$ $-7 - 0 = -7$ $12 - +4 = 8$
$9 + -5 = 4$ $-7 - -2 = -5$ $8 - 0 = 8$
$9 + -10 = \square$ $-7 - -4 = \square$ $4 - -4 = \square$
$9 + \square = \square$ $-7 - \square = \square$ $0 - \square = \square$
$9 + \square = \square$ $-7 - \square = \square$ $\square - \square = \square$

8 Work out the answers.

a $13 - 12$ **b** $-24 - -13$ **c** $17 - -26$ **d** $-71 + -3$ **e** $37 - 13$
f $-29 - -15$ **g** $-16 + 16$ **h** $26 - -9$ **i** $-9 + -9$ **j** $-11 + -18$
k $15 - 17$ **l** $47 - -25$ **m** $-24 - -32 + -41$ **n** $-13 + 12 - - 24$ **o** $-103 + 75$

9 These temperatures were recorded at an airport in January.

Copy and complete the table. Draw a number line to check your answers.

	Sunday	Monday	Tuesday	Wednesday	Thursday	Friday	Saturday
Maximum temperature (°C)	8	5	−1		2	2	9
Minimum temperature (°C)	−11	−6		−9	−8		−2
Difference (degrees)	19		8	14		6	

(MR) **10** A fish is 12 m below the surface of the water. A fish eagle is 17 m above the water. How many metres must the bird descend to get the fish?

Challenge: Marking a test

A A test consists of 50 questions. A correct answer earns 3 marks, a wrong answer gets −2 marks and −1 mark is given if an answer is not attempted.

Work out each pupil's score.

a Eve gets 32 right, 10 wrong and did not attempt 8.

b Sophia gets 20 right, 20 wrong and did not attempt 10.

c Oliver gets 25 right and 25 wrong.

B Andrew scores 104 points. Can you work out how he did this?

A computer spreadsheet is useful for this activity.

1.5 Multiplying negative numbers

Learning objective

• To carry out multiplications involving negative numbers

Key words

| multiply | product |

This diagram shows the result of multiplying both positive and negative numbers by a positive number. In this example all numbers are multiplied by +2.

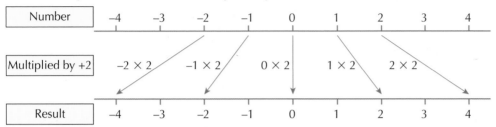

This shows that:

• multiplying a positive number by another positive number gives a positive number

• multiplying a negative number by a positive number gives a negative number.

To summarise this:

$(-) \times (+) = (-)$ and $(+) \times (+) = (+)$

Example 11

Work out the answers.

a -12×4 **b** -7×3

a $-12 \times 4 = -48$ **b** $-7 \times 3 = -21$

What happens if you multiply a number by a negative number?

This diagram shows positive and negative numbers multiplied by −2.

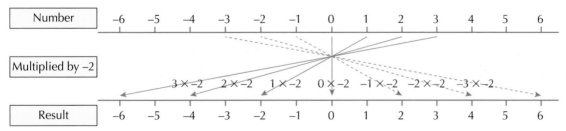

This shows that a positive number multiplied by a negative number gives a negative result, as in the first diagram. Here it is just shown the other way round. But this diagram also shows that a negative number multiplied by a negative number gives a positive number.

To summarise this:

$(-) \times (-) = (+)$ and $(-) \times (+) = (-)$

To help you remember:

- When multiplying numbers with a different sign, the answer is negative.
- When multiplying numbers with the same sign, the answer is positive.

Example 12

Work out the answers.

a $8 \times (-5)$ **b** -4×-6

   a $8 \times (-5) = -40$ **b** $-4 \times -6 = 24$

Exercise 1E

1 Work out the answers.

   a $-2 \times (-1)$ **b** $-3 \times (-4)$ **c** -6×7 **d** -1×2

   e $-4 \times (-3)$ **f** $8 \times (-2)$ **g** $5 \times (-2)$ **h** $-2 \times (-6)$

   i -3×1 **j** -4×10 **k** $-3 \times (-3)$ **l** $(-8) \times (-9)$

2 Work out the answers.

   a -2×-7 **b** -3×-9 **c** -7×8 **d** -1×12

   e -3×-5 **f** 8×-6 **g** 4×-11 **h** -2×-8

   i -5×1 **j** -3×13 **k** -8×-12 **l** -7×-9

3 In each of the brick walls below, you need to work out the number to write in an empty brick by multiplying the numbers in the two bricks below it. Copy and complete each brick wall.

a

-4	

4	-1	2

b

-3	4	5

c

-2	-4	3

PS **4** **a** Julie asked Chris to think of two numbers smaller than ten and tell her their product.

The product is –24.

There are four different possible sets of numbers that give that product.

Write down the four possible pairs of numbers Chris could have been thinking of.

b Chris asked Julie to think of two numbers smaller than ten and tell him their product.

Julie said: 'The product is 12.'

Chris said that there were four different possible sets of numbers with that product.

Write down the four possible pairs of numbers Julie could have been thinking of.

5 Copy and complete the following multiplication grids.

a

×	1	2	3	4
−5				
−6				
−7				
−8				

b

×	−1	−3	−5	−7
−9				
−11				
−13				
−15				

6 Work out the answers.

a 7×-18 **b** $3 - 12$ **c** 11×-14 **d** $2 - -18$ **e** -9×13 **f** $-33 - 33$

g $-10 + 23$ **h** -8×-11 **i** -11×-12 **j** $-15 + 7$ **k** $-11 - 15$ **l** -14×9

m -31×-4 **n** $-7 + 23$ **o** -25×12 **p** $-12 + 13$ **q** $-7 + 18$ **r** -8×-18

7 Write down the next three numbers in each number sequence.

a 1, −2, 4, −8, …, …, … **b** −1, −3, −9, −27, …, …, …

c −1, 5, −25, 125, …, …, … **d** 1, −4, 16, −64, …, …, …

8 a In each brick wall, work out the number to write in an empty brick by multiplying the numbers in the two bricks below it. Copy and complete each brick wall.

i **ii**

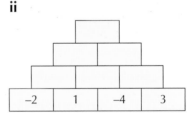

b Andy said: 'You will always have a positive number at the top of the brick wall if there are two negative numbers in the bottom layer.'

Is Andy correct? Explain your answer.

c What combination of positive and negative numbers do you need on the bottom layer to end up with a negative number at the top?

Challenge: Number puzzle

A Choose any negative number from −1 to −12.

Subtract 9, then multiply by −2 and add the number you first thought of.

Now add together the digits of the final number.

What do you notice?

B Try this again with more numbers.

Ready to progress?

I can show my understanding of how the number line extends to include negative numbers.
I can understand and use the inequality symbols < for less than and > for greater than.
I can use my understanding of negative numbers to solve simple real-life problems.
I can carry out addition and subtraction involving negative numbers.

I can carry out multiplications involving negative numbers.

Review questions

1   The keel (bottom) of a cruise ship is 7 m below the surface of the water. The deck is 27 m above the water.

What is the total height, from deck to keel?

(PS) 2   Harry is on the 29th floor of a skyscraper. He goes up 18 floors and then down 23 floors. He wants to go to the 2nd floor.

How many floors does he need to go down now?

(FS) 3   a   Alisha opens a bank account with a deposit of £400. In the next two weeks she takes out £170, deposits £130 and takes out £120.

How much is in her account now?

b   Next, Alisha deposits £60 and takes out £170.
How much is in her account now?

(PS) 4   Karen and Geza have parked their car on level −5 of the shopping centre car park.

They take the lift to the shops on level +6, then Karen realises she has left her purse in the car. She goes back to the car to get it. Then she returns to the shops and meets Geza on level +4.

How many levels has Karen travelled through altogether?

5   This is a bank statement.

Statement number: 10				Account number 14628769
Date	Details	Debit (£)	Credit (£)	Balance (£)
31-01-2014	Opening balance			827.54
01-02-2014	Wage		252.71	
03-02-2014	Cash withdrawal	130.00		
05-02-2014	Transfer to savings account	250.00		
08-02-2014	Wage		252.71	
28-02-2014	Closing balance			

a Copy the statement and complete the balance column.

b Explain the difference between a debit and a credit.

6 The Mariana Trench in the Pacific Ocean has a maximum depth of 10911 metres below sea level. The summit of Mount Everest is 8848 metres above sea level.

a How much higher than the base of the trench is the summit of Mount Everest?

b If Mount Everest was set in the deepest part of the trench, how far would its summit be below sea level?

7 In a popular TV programme, each of two teams has to buy three items at an antiques fair.

The items are sold at auction and the team that makes more money wins and keeps any profit they make.

Copy and complete each team's score sheet.

a Red team

Item	Buying price (£)	Selling price (£)	Profit (£)
Silver dish	59		−18
Umbrella		47	+14
Toy car	45	55	
Total	137		

b Blue team

Item	Buying price (£)	Selling price (£)	Profit (£)
Necklace		55	+17
Doll	49	85	
Watch	110		−52
Total	197		

c Which team won and by how much?

d Teams can go for a bonus buy, chosen by an expert.

The Red team bonus buy was bought for £33 and sold for £25.

The Blue team bonus buy was bought for £21 and sold for £29.

Copy and complete this table to show who would win in each case.

	Red team without the bonus buy	Red team with the bonus buy
Blue team without the bonus buy		
Blue team with the bonus buy		

8 Work out the answers.

a $5 \times (-1)$ b -3×9 c -8×-7 d -11×12

e $-9 \times (-7)$ f $8 \times (-3)$ g $5 \times (-12)$ h $-12 \times (-8)$

i -23×-1 j -14×100 k $-30 \times (-10)$ l $(-8) \times (-90)$

Problem solving
Where in the world?

A Where shall we go?

David and Hannah were discussing where they could go for their honeymoon in the summer.

They were interested in the places listed in the table below.

They worked out the time difference from the UK for each one.

1   Which place is furthest behind the UK, in terms of time?

2   What is the time difference between Reykjavik and Amsterdam?

3   What is the time difference between San Francisco and New York?

4   What time will it be in Bangkok when it is 10:00 am in the UK?

5   What time is it in New Delhi when it is 6:00 pm in New York?

City	Difference from UK time (hours)
Amsterdam	+1
Bangkok	+7
Hong Kong	+8
New Delhi	+5:30
New York	−5
Reykjavik	0
San Francisco	−8
Sydney	+11

B How shall we get there?

The couple decided to go to Sydney, so they looked up some flight times and prices.

David and Hannah cannot get a direct flight from Heathrow to Sydney.

1 What is the difference in price between the most expensive flight and the cheapest flight to Sydney?

2 They plan to leave Heathrow on 20 August.
What date and time would each flight get them into Sydney?

3 They need to arrive back into Heathrow on 3 September.
What date and times could they get a flight back?

4 They want to leave Heathrow on 20 August and be back in Heathrow on 3 September.
What combination of flights gives them:

   a the longest time in Sydney

   b the least time in Sydney?

	Heathrow to Sydney			Sydney to Heathrow		
Fare (£)	Departure	Arrival (local time)	Approximate journey time (hours)	Departure	Arrival (local time)	Approximate journey time (hours)
830	21:00	10:00	26	09:30	15:30	41
890	16:00	18:00	39	10:50	05:50	30
1092	19:15	07:15	25	14:00	03:00	24

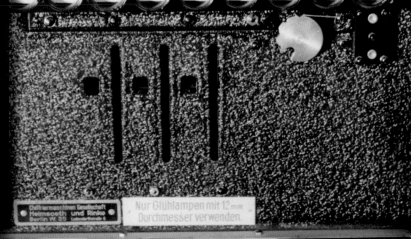

2

Sequences

This chapter is going to show you:

- how to use function machines to generate input or output values
- how to describe some simple number patterns
- how to create sequences and describe them in words
- how to generate and describe sequences that include fractions and decimals
- how to work out and use the *n*th term of sequences
- how to use the special sequence called the sequence of square numbers
- how to use the special sequence called the sequence of triangular numbers
- how to use the Fibonacci sequence and Pascal's triangle.

You should already know:

- odd and even numbers
- multiplication tables up to 12×12
- how to apply the four rules of number.

About this chapter

During the Second World War the first computer in the world was invented at Bletchley Park in the UK. At that time, Britain was at war with Germany and needed to break the coded German communications to discover what they were planning to do next. Codes are based on sequences and these were very complex ones, which were changed every day and randomly generated by a machine called Enigma. It was the job of the computer to crack each day's new code sequences from the Enigma machine – and fast. Today, coded sequences are still used in secure communications, for example, encrypting websites used for financial transactions – vital to everyday business.

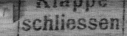

2.1 Function machines

Learning objectives

- To use function machines to generate inputs and outputs
- To use given inputs and outputs to work out a function

In mathematics, a **function** is an **operation**, or a set of operations, that changes one number into another. There are many ways to show functions, one of which is the **function machine**. The numbers you start with are called the **input**. The numbers you get after you apply the operations are the **output**.

The operations in a function machine are called its **rules**.

Function machines are useful because they help you to understand the rules of algebra when you are solving equations.

Example 1

Complete a function machine to show the output for the numbers 3, 5 and 7 for the rule: multiply by 3 and then subtract 10.

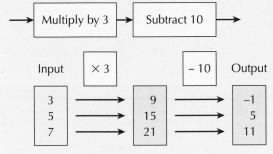

So the outputs are: −1, 5 and 11.

If you knew that the output from the function machine in Example 1 was 26, how could you work out the input?

You would work backwards by working out:

$$26 + 10 = 36$$

$$36 \div 3 = 12$$

So the input would be 12.

One way of doing this is to create an **inverse** function machine.

The inverse operation for add is subtract and vice versa and the inverse operation for multiplication is divide and vice versa.

The reverse function machine for Example 1 looks like this.

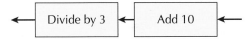

Example 2

a Draw an inverse function machine for this function machine.

b Work out the output for an input of 5 to the function machine.

c Put your answer for part **b** through the inverse function machine.

d Work out an input that gives the same value for the output.

a

b $5 \times 5 = 25, 25 - 8 = 17$

c $17 + 8 = 25, 25 \div 5 = 5$

d Guess an input and see if the outputs are the same. Then keep on guessing until you get the right answer. If the input is 2, then the output is $2 \times 5 = 10, 10 - 8 = 2$.

This answer will work for the original function machine and the inverse function machine.

Exercise 2A

1 Write down the rule for each set of inputs and outputs, in words.

a

Input	Output
5 →	9
8 →	12
11 →	15
−3 →	1

b

Input	Output
6 →	18
8 →	24
10 →	30
21 →	63

c

Input	Output
$2\frac{1}{2}$ →	10
6 →	24
11 →	44
$12\frac{1}{2}$ →	50

d

Input	Output
3 →	$1\frac{1}{2}$
8 →	4
18 →	9
21 →	$10\frac{1}{2}$

e

Input	Output
5 →	−9
7 →	−7
9 →	−5
15 →	1

f

Input	Output
2 →	−2
4 →	−4
6 →	−6
8 →	−8

2 Work out the missing numbers in these function machines.

a

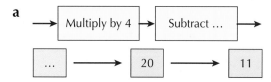

b

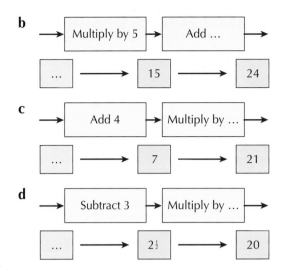

c

d

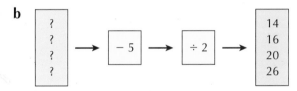

PS **3** Use inverse operations to work out the inputs for each of these function machines.

a

?				7
?	→	× 2 →	+ 3 →	13
?				16
?				25

b

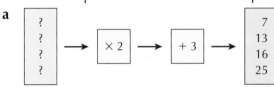

?				14
?	→	− 5 →	÷ 2 →	16
?				20
?				26

c

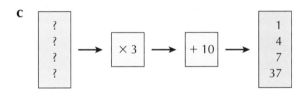

?				1
?	→	× 3 →	+ 10 →	4
?				7
?				37

PS **4** The input and output for this function machine are both 10.

Work out possible values for the operations shown in the boxes.

| 10 | → | Multiply by ... | → | Subtract ... | → | 10 |

5 Each of these function machines uses two operations.

Work out what they are, for each one.

a

Input		Output
1	→	4
2	→	9
3	→	14
4	→	19

b

Input		Output
1	→	10
2	→	15
3	→	20
4	→	25

c

Input		Output
1	→	2
2	→	$2\frac{1}{2}$
3	→	3
4	→	$3\frac{1}{2}$

6 Draw an inverse function machine for each of these function machines.

Use the inverse function machines to work out the input for an output of 42.

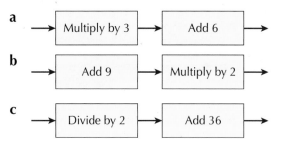

a → Multiply by 3 → Add 6 →

b → Add 9 → Multiply by 2 →

c → Divide by 2 → Add 36 →

d What do you notice about your answers?

PS **7** Work out the single input that gives the same output from both of these function machines.

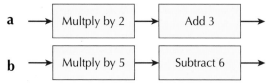

a → Multply by 2 → Add 3 →

b → Multply by 5 → Subtract 6 →

PS **8** Work out the single input that gives the same output from both of these function machines.

a → Divide by 3 → Add 2 →

b → Add 6 → Multiply by 2 →

Challenge: Inputs and outputs

A Look at each number machine and work out the input that will give the same output as the input.

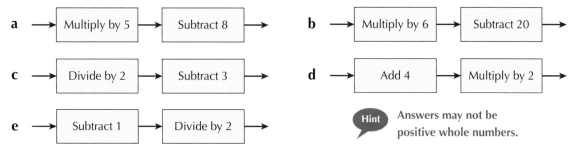

a → Multiply by 5 → Subtract 8 →   **b** → Multiply by 6 → Subtract 20 →

c → Divide by 2 → Subtract 3 →   **d** → Add 4 → Multiply by 2 →

e → Subtract 1 → Divide by 2 →

Hint Answers may not be positive whole numbers.

B These are the incomplete rules for a function machine.

→ Multiply by ... → Add ... →

a Work out numbers to go in the rule boxes so that:

   i input 5 → output 17   **ii** input 8 → output 60   **iii** input 13 → output 65.

b Look back at part **a**. In each example you could put more than one pair of numbers into the rule boxes to give those inputs and outputs. See how many pairs you can find.

2.2 Sequences and rules

Learning objective

- To recognise, describe and generate sequences that follow a simple rule

Key words

first term	geometric sequence
linear sequence	sequence
term	term-to-term rule

A **sequence** is a list of numbers that follow a pattern or rule. You can use simple rules to make up many different sequences with whole numbers. Sequences may also have different starting points. With *different* rules and *different* starting points, there are very many *different* sequences you may make.

Example 3

Rule: add 3   Starting at 1 gives the sequence 1, 4, 7, 10, 13, …

Starting at 2 gives the sequence 2, 5, 8, 11, 14, …

Starting at 6 gives the sequence 6, 9, 12, 15, 18, …

Example 4

Rule: double   Starting at 1 gives the sequence 1, 2, 4, 8, 16, …

Starting at 3 gives the sequence 3, 6, 12, 24, 48, …

Starting at 5 gives the sequence 5, 10, 20, 40, 80, …

The numbers in a sequence are called **terms** and the starting number is called the **first term**. The rule is often called the **term-to-term rule**.

Sequences that increase or decrease by a fixed amount from one term to the next are called **linear sequences**. Example 3 above shows linear sequences.

Sequences where each term after the first is found by multiplying or dividing by a fixed amount are called **geometric sequences**. Example 4 above shows geometric sequences.

Exercise 2B

1 Use each term-to-term rule to write down the first five terms of a sequence.

Start from a first term of 1.

    a add 6   **b** multiply by 9   **c** subtract 3   **d** divide by 2

2 Use each term-to-term rule in question 1, starting from a first term of 5, to write down the first five terms of a new sequence.

3 Work out the next two terms in each sequence.

Describe the term-to-term rule you have used.

    a 3, 7, 11, …, …   **b** 4, 16, 64, …, …   **c** 1, 5, 25, …, …

    d 8, 16, 32, …, …   **e** 100, 50, 25, …, …   **f** 9, 5, 1, …, …

    g 1000, 200, 40, …, …   **h** 2, 6, 18, …, …

4 Work out the next two terms in each sequence.

Describe the term-to-term rule you have used.

a 20, 15, 10, 5, 0, …, …
b 4, 2, 1, $\frac{1}{2}$, $\frac{1}{4}$, …, …
c 66, 50, 34, 18, 2, …, …
d 1000, 100, 10, 1, 0.1, …, …

5 For each pair of numbers, work out at least two different sequences and write down the next two terms.

Describe the term-to-term rule you have used.

a 1, 5, …, …
b 3, 6, …, …
c 2, 8, …, …
d 3, 12, …, …
e 4, 2, …, …
f 45, 15, …, …

PS **6** Work out two terms between each pair of numbers, to form a sequence.

Describe the term-to-term rule you have used.

a 1, …, …, 10
b 18, …, …, 3
c 5, …, …, 625
d 4, …, …, $\frac{1}{16}$
e 160, …, …, 20
f 3, …, …, 81

MR **7** A snail is at the bottom of a well that is 12 feet deep. During the day the snail climbs 3 feet up the well, but during the night it slides 2 feet back down again.

How many days will it take the snail to reach the top of the well?

MR **8** Six fence posts are fixed, at equal distances apart, in a straight line. The distance between the two end posts is 12 feet.

What distance is each post in the fence from the next one?

FS **9** Laura saves money for 10 weeks, starting with £5 in the first week, £10 the second week, £15 the third week and so on.

Ed saves money for 10 weeks, starting with 50p the first week, £1 the second week, £2 the third week and so on.

Who has more money after 10 weeks, and by how much?

Investigation: Common terms

A **a** Write down the first 20 terms of the sequence that starts 5, 9, 13, 17, 21, … .

b Write down the first 20 terms of the sequence that starts 3, 8, 13, 18, 23, … .

B 13 is a common term in both sequences.

a What are the other common terms?

b Is there a rule to them?

 Hint You could use a spreadsheet to do this investigation.

C Here are another two sequences.

1, 5, 9, 13, 17, …

2, 8, 14, 20, 26, …

Are there any common terms in both sequences?

Explain your answer.

2.3 Working out missing terms

Learning objective

• To work out missing terms in a sequence

The terms in a sequence are described as the first term, second term, third term, fourth term and so on.

You need to know how to work out any term in a sequence.

Example 5

Look at this sequence:

7, 10, 13, 16, …

What is the 5th term? What is the 25th term? What is the 50th term?

You first need to know what the term-to-term rule is.

You can see that you add 3 from one term to the next.

This is called an 'add 3' rule.

```
     + 3    + 3    + 3
7       10     13     16      …
1st     2nd    3rd    4th     5th  …  50th
```

To get to the fifth term, you add 3 to the fourth term, which gives 19.

To get to the 25th term, you will have to add on 3 a total of 24 times (25 − 1) to the first term, 7.

This will give $7 + 3 \times 24 = 7 + 72 = 79$.

To get to the 50th term, you will have to add on 3 a total of 49 times (50 − 1) to the first term, 7.

This will give $7 + 3 \times 49 = 7 + 147 = 154$.

Example 6

Look at this sequence:

45, 40, 35, 30, …

What is the 5th term? What is the 50th term?

```
     − 5    − 5    − 5
45      40     35     30      …
1st     2nd    3rd    4th     5th  …  50th
```

This is a 'subtract 5' rule. To get to the fifth term, you subtract 5 from the fourth term, which gives 25.

To get to the 50th term, you will have to subtract 5 a total of 49 times (50 − 1) from the first term, 45.

This will give $45 − 5 \times 49 = 45 − 245 = −200$.

1 Work out the fifth term and the 50th term in each sequence.

    a   7, 13, 19, 25, …   **b**   2, 7, 12, 17, …   **c**   7, 14, 21, 28, …

    d   10, 18, 26, 34, …   **e**   1, 9, 17, 25, …   **f**   17, 27, 37, 47, …

PS **2** In each sequence, you have been given the fourth, fifth and sixth terms.

Work out the first term and the 50th term in each case.

    a   …, …, …, 5, 7, 9, …   **b**   …, …, …, 7, 12, 17, …

    c   …, …, …, 9, 13, 17, …   **d**   …, …, …, 2, 11, 20, …

3 Work out the 20th term in each sequence.

    a   98, 95, 92, 89, 86, ….   **b**   57, 50, 43, 36, 29, ….

    c   42, 38, 34, 30, 26, …..   **d**   38, 31, 24, 17, 10, …

4 Work out the missing terms and the 50th term in each sequence.

Term	1st	2nd	3rd	4th	5th	6th	7th	8th	50th
Sequence A	…	…	…	…	12	16	20	24	…
Sequence B	…	2	…	12	…	22	…	32	…
Sequence C	…	…	9	16	…	30	37	…	…
Sequence D	…	…	11	…	33	…	…	66	…

5 Work out the 30th term in the sequence with the term-to-term rule 'add 7' and a first term of 5.

6 Work out the 20th term in the sequence with the term-to-term rule 'subtract 5' and a first term of 94.

PS **7** These patterns are made from mauve and white squares.

The diagrams show the patterns for the third and fifth terms.

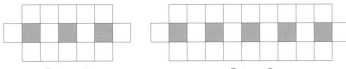

    Pattern 3   Pattern 5

    a   How many mauve squares are there in the pattern for the fourth term?

    b   How many white squares are there in the pattern for the fourth term?

    c   Draw the pattern for the first term.

    d   How many squares in total are there in the pattern for the 10th term?

PS **8** The second and third terms of a sequence are 3 and 6.

…, 3, 6, …

There are several different sequences that could have 3 and 6 as the second and third terms.

    a   Write down one rule for the sequence and work out the first and fourth terms.

    b   Write down a different rule for the sequence and work out the first and fourth terms.

a What is the term-to-term rule here? It is add 4, so the rule is based on $4n$.

The first term is 5.

For the first term, $n = 1$ and $4 \times 1 = 4$, but the first term is 5, so you need to add 1 to $4n$.

So the nth term is $4n + 1$.

b Now use this rule to work out the 50th term in the pattern.

When $n = 50$, $4n + 1 = 4 \times 50 + 1 = 201$.

Example 10

Work out the nth term of the sequence 2, 5, 8, 11, 14, …

This is an 'add 3' rule so the rule is based on $3n$.

The first term is 2, so subtract 1 from $3n$.

Hence the nth term is $3n - 1$.

Exercise 2D

1 Work out:

   i   the first three terms   **ii**   the 100th term

   for the sequence, if the nth term is:

  a $2n + 1$ **b** $4n - 1$ **c** $5n - 3$

  d $3n + 2$ **e** $4n + 5$ **f** $10n + 1$

  g $\frac{1}{2}n + 2$ **h** $7n - 1$ **i** $\frac{1}{2}n - \frac{1}{4}$.

2 For each of the patterns below, work out:

   i   the nth term for the number of matchsticks

   ii the number of matchsticks in the 50th term.

a

b

c

d

3 Work out the nth term of each sequence.

  a 3, 9, 15, 21, 27, …   **b** 10, 13, 16, 19, 22, …   **c** 7, 13, 19, 25, 31, …

  d 1, 4, 7, 10, 13, …   **e** 4, 11, 18, 25, 32, …   **f** 5, 7, 9, 11, 13, …

  g 9, 13, 17, 21, 25, …   **h** 5, 13, 21, 29, 37, …   **i** 11, 21, 31, 41, 51, …

  j 3, 12, 21, 30, 39, …

4 Each pattern below is made from matchsticks of two different colours.

Work out the *n*th term for:

i the number of red-tipped matchsticks

ii the number of blue-tipped matchsticks

iii the total number of matchsticks.

Use your rules to describe the 50th term in the patterns by working out:

iv the number of red-tipped matchsticks

v the number of blue-tipped matchsticks

vi the total number of matchsticks.

a

b

5 Work out the *n*th term of each sequence.

a 95, 90, 85, 80, 75, ...   **b** 33, 26, 19, 12, 5, ...

c 24, 21, 18, 15, 12, ...   **d** 10, 8, 6, 4, 2 ...

6 Work out the *n*th term of each sequence. They may include decimals or fractions.

a $2\frac{1}{2}$, 4, $5\frac{1}{2}$, 7, $8\frac{1}{2}$, ...   **b** 2.8, 4.1, 5.4, 6.7, 8.0, ...

c 7, $9\frac{1}{2}$, 12, $14\frac{1}{2}$, 17, ...   **d** 32.5, 30, 27.5, 25, 22.5, ...

7 Work out the first five terms of the sequence, if the *n*th term is:

a n^2 **b** $n^2 + 4$ **c** $n^2 - 1$

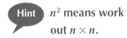

Hint n^2 means work out $n \times n$.

MR **8** What are the differences between the terms in each of the three sequences in question **7**? What do you notice?

Investigation: The eccentric mathematician

The daughter of a rich but eccentric mathematician had twin girls, Ellen and Emily.

The mathematician started a savings account for each of his granddaughters.

In one account he put in £5000 each year.

In the other account he put in £2 the first year, £4 the second year, £8 the third year and kept on putting in double the amount each year.

When the twins were 10 years old he asked them to choose which account they wanted.

A Show, by setting up a table or using a spreadsheet, that the first account had £50 000 in it and the second account had £2046.

B Ellen immediately chose the first account. Emily was left with the second account.

The grandfather kept on paying in money until the twins were 18 years old.

Which twin had more money in her account at age 18?

2.5 Other sequences

Learning objective

- To know and understand the square and triangular number sequences, the Fibonacci sequence and Pascal's triangle

Key words

Fibonacci sequence	Pascal's triangle
power	square numbers
squaring	triangular numbers

Square numbers and triangular numbers

When you multiply a number by itself you are **squaring** the number. The result is a **square number**. For example:

- 4 is a square number and it is the square of 2 ($2 \times 2 = 4$)
- 9 is a square number and it is the square of 3 ($3 \times 3 = 9$) and so on.

Instead of writing 1×1, 2×2, 3×3 and so on, you can write 1^2, 2^2, 3^2. This small 2 is called a **power** and, because it is so special, the power of 2 is also called 'square'. You refer to the number it is attached to as 'squared', for example, you say 5^2 as 'five squared'.

This table shows the first 10 square numbers. You can see from the bottom line of the table why they are called square numbers.

1×1	2×2	3×3	4×4	5×5	6×6	7×7	8×8	9×9	10×10
1^2	2^2	3^2	4^2	5^2	6^2	7^2	8^2	9^2	10^2
1	4	9	16	25	36	49	64	81	100

This is the start of the sequence of square numbers. You need to learn the square numbers up to $15^2 = 225$.

Another well-know sequence is 1, 3, 6, 10, 15, 21, ….

This is called the sequence of **triangular numbers**. This sequence builds up by adding on one more each time.

First term: 1

Second term: add 2 to the first term ($1 + 2 = 3$)

Third term: add 3 to the second term ($3 + 3 = 6$)

Fourth term: add 4 to the third term ($6 + 4 = 10$)

Fifth term: add 5 to the fourth term ($10 + 5 = 15$)

This table shows you the first 10 triangular numbers. You can see from the bottom line of the table why they are called triangular numbers.

1	$1 + 2$	$3 + 3$	$6 + 4$	$10 + 5$	$15 + 6$	$21 + 7$	$28 + 8$	$36 + 9$	$45 + 10$
1	3	6	10	15	21	28	36	45	55

You will find these numbers used in ten-pin bowling and snooker.

Exercise 2E

1 Copy the first three rows of the table of square numbers and continue it up to 15×15.

2 Write each number below as the sum of two square numbers. The first two have been done for you.

a $5 = 1 + 4$ **b** $10 = 1 + 9$ **c** $13 = \ldots + \ldots$

d $17 = \ldots + \ldots$ **e** $20 = \ldots.. + \ldots$ **f** $25 = \ldots + \ldots$

g $26 = \ldots + \ldots$ **h** $29 = \ldots.. + \ldots$ **i** $34 = \ldots + \ldots$

j $37 = \ldots + \ldots$ **k** $40 = \ldots.. + \ldots$ **l** $41 = \ldots + \ldots$

m $45 = \ldots + \ldots$ **n** $50 = \ldots.. + \ldots$ **o** $52 = \ldots + \ldots$

3 Copy the first two rows of the table of triangular numbers and continue it up to the 15th triangular number.

4 Write each number as the sum of two triangular numbers. The first two have been done for you.

a $4 = 1 + 3$ **b** $7 = 1 + 6$ **c** $9 = \ldots + \ldots$

d $11 = \ldots + \ldots$ **e** $13 = \ldots + \ldots$ **f** $16 = \ldots + \ldots$

g $18 = \ldots + \ldots$ **h** $21 = \ldots + \ldots$ **i** $22 = \ldots + \ldots$

j $24 = \ldots + \ldots$ **k** $25 = \ldots + \ldots$ **l** $27 = \ldots + \ldots$

m $29 = \ldots + \ldots$ **n** $31 = \ldots + \ldots$ **o** $34 = \ldots + \ldots$

5 Write down two numbers that are both square numbers and also triangular numbers.

6 Look at this pattern of numbers.

$$1 \qquad\qquad\quad = \ 1 \ = 1^2$$

$$1 + 3 \qquad\qquad = \ 4 \ = 2^2$$

$$1 + 3 + 5 \qquad\ \ = \ 9 \ = 3^2$$

$$1 + 3 + 5 + 7 \quad\ = 16 \ = 4^2$$

$$1 + 3 + 5 + 7 + 9 = 25 \ = 5^2$$

a Write down the next two lines of this number pattern.

b What is special about the numbers on the left hand side?

(PS) **c** Without working them out, write down the answers to these calculations.

    i $1 + 3 + 5 + 7 + 9 + 11 + 13 + 15 + 17 + 19 = \ldots$

    ii $1 + 3 + 5 + 7 + 9 + 11 + 13 + 15 + 17 + 19 + 21 + 23 + 25 + 27 + 29 = \ldots$

7 Some sums of two square numbers are special because they give an answer that is also a square number. For example:

$3^2 + 4^2 = 9 + 16 = 25 = 5^2$

Which of these pairs of square numbers give a total that is also a square number?

a $5^2 + 12^2$ **b** $2^2 + 5^2$ **c** $6^2 + 8^2$

d $7^2 + 9^2$ **e** $7^2 + 24^2$ **f** $10^2 + 24^2$

8 **a** Add up the first 10 pairs of consecutive square numbers, starting with 1 + 4, 4 + 9, 9 + 16, … .

b Is it possible to get an even total if you add any pair of consecutive square numbers? If not, explain why not.

c Work out the differences between the totals.

Do they form a sequence? If so, describe the sequence.

9 Look at this pattern of numbers.

$$1 = 1$$
$$1 + 2 = 3$$
$$1 + 2 + 3 = 6$$
$$1 + 2 + 3 + 4 = 10$$
$$1 + 2 + 3 + 4 + 5 = 15$$

a Write down the next two lines of this number pattern.

b What is special about the numbers on the left-hand side?

c What is special about the numbers on the right-hand side?

d Without working them out, write down the answers to these calculations.

i $1 + 2 + 3 + 4 + 5 + 6 + 7 + 8 + 9 + 10 = \ldots$

ii $1 + 2 + 3 + 4 + 5 + 6 + 7 + 8 + 9 + 10 + 11 + 12 + 13 + 14 + 15 = \ldots$

10 **a** Add up the first 10 pairs of consecutive triangular numbers, starting with 1 + 3, 3 + 6, 6 + 10, ….

b What is special about the answers?

Challenge: Testing rules

A Here is a rule.

8 × any triangular number + 1 is a square number.

Test this rule for the first triangular number 1.

$$8 \times 1 + 1 = 8 + 1 = 9 = 3^2$$

Does the rule always work?

Test it using at least four other triangular numbers.

B Now test this rule.

9 × any triangular number + 1 is also a triangular number.

C Now test this rule.

The sum of the squares of any two consecutive triangular numbers is also a triangular number.

Special sequences

The next two number patterns are not linear or geometric sequences but they are very important in mathematics.

The Fibonacci sequence can also be found in many patterns in nature.

The Fibonacci sequence

An Italian mathematician, Leonardo Fibonacci first wrote about this sequence in the 13th century. The sequence is:

1   1   2   3   5   8   13   21   ...

Each term after the first two is the sum of the previous two terms.

The next term would be $13 + 21 = 34$, and so on.

Pascal's triangle

A French mathematician, Blaise Pascal, first wrote about this number pattern, which is now known as Pascal's triangle.

Row

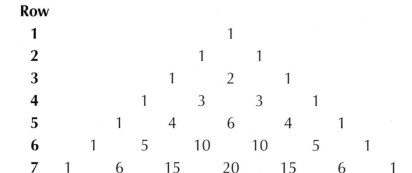

Row								
1					1			
2				1		1		
3			1		2		1	
4		1		3		3		1
5	1		4		6		4		1
6	1		5		10		10		5		1
7	1		6		15		20		15		6		1

Each row starts and finishes with 1.

Each of the other numbers is the sum of the two numbers above it, to the left and right.

For example, in row 5:

4 = 1 + 3   6 = 3 + 3

and in row 7:

15 = 10 + 5.

Exercise 2F

1. Continue the Fibonacci sequence until the 20th term.

2. Write down four numbers that are in both the sequence of triangular numbers and the Fibonacci sequence.

3. Write down two numbers that are in both the sequence of square numbers and the Fibonacci sequence.

4. Look again at Pascal's triangle, above, and write down the next three rows.

(MR) 5. Add up the numbers in each row of Pascal's triangle:

1 = 1, 1 + 1 = 2, 1 + 2 + 1 = 4, ...

Describe the number pattern formed by the totals.

6 Look again at Pascal's triangle.

    a Find a diagonal that gives the counting numbers (1, 2, 3, 4, …).

    b Find a diagonal that gives the triangular numbers.

 7 Add up the first three numbers in the Fibonacci sequence, then the first four, then the first five, and so on. Can you see any pattern emerging in the results?

 8 Write down any four consecutive numbers from the Fibonacci sequence, for example, 5, 8, 13, 21.

        Multiply the first and the last together.

        Multiply the middle two together.

        What is the difference between the numbers?

        Try this for at least another two sets of four consecutive Fibonacci numbers.

        Does this rule work for any four consecutive Fibonacci numbers?

Investigation: Consecutive numbers

A Take any three consecutive numbers, for example, 7, 8, 9.

Multiply the first and last of your numbers. $7 \times 9 = 63$

Square the middle number. $8^2 = 64$

Subtract the first number from the second. $64 - 63 = 1$

Try this with at least three different sets of consecutive numbers.

What happens in every case?

B Now try the same process as in part **A** with any three consecutive Fibonacci numbers, For example 5, 8, 13.

Repeat this with at least three different sets of consecutive Fibonacci numbers.

What happens this time?

C Now try the same process as in part **A** with any three alternate consecutive Fibonacci numbers, for example, 3, 8, 21.

Repeat this with at least three different sets of alternate consecutive Fibonacci numbers.

What happens this time?

D Now try the same process as in part **A** with any three Fibonacci numbers that are consecutive but with a gap of two numbers between them, for example, 2, 8, 34.

Repeat this with at least three different similar sets of Fibonacci numbers.

What happens this time? Can you write down a rule?

Ready to progress?

I can work out the term-to-term rule for a sequence.
I can work out the operation in a function machine that uses one rule, when I am given the inputs and outputs.
I know the square numbers up to 15×15.

I can work out the inputs when given the outputs and the rules for an inverse function machine.
I can work out any term in a sequence, given the first term and the term-to-term rule.
I can recognise and work out the sequence of triangular numbers.
I can investigate the patterns and connections within the square and triangular numbers.
I can recognise and work out the numbers in the Fibonacci sequence and in Pascal's triangle.
I can work out any term in a sequence, given the nth term.

I can work out the operations in a function machine that uses more than one rule, when I am given the input and output values.
I can work out the nth term for any linear sequence.
I can investigate a given rule and reach a conclusion about whether it always works.

Review questions

1   Here is a sequence of shapes made with blue and white tiles.

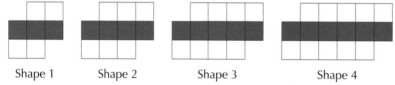

Shape 1   Shape 2   Shape 3   Shape 4

The number of blue tiles   =   the shape number + 2
The number of white tiles   =   2 × the shape number + 2

a   How many blue tiles will there be in shape number 8?

b   How many white tiles will there be in shape number 35?

c   How many tiles altogether will there be in shape number 50?

d   Write down the missing numbers from this sentence:
The total number of tiles = … × the shape number + …

2   In each of these function machines, the two rules can be replaced with one rule.

Work out what this is for each machine. The first one has been done for you.

a   → +2 → −7 → = → −5 →   b   → −7 → +3 → = → ? →

c   → ×9 → ×2 → = → ? →   d   → ×4 → ÷8 → = → ? →

3 **a** Write down the next two numbers in the sequence below.

93, 85, 77, 69, 61, …, …

b Write down the next two numbers in this sequence.

1, 2, 4, 7, 11, 16, …, …

c What two numbers do both sequences have in common, if they are continued?

4 **a** Maisy saves £5 each week for 10 weeks. How much will she have after five weeks?

b Daisy saves £1 the first week, £2 the second week, £3 the third week, £4 the fourth week, and so on for 10 weeks. Copy and complete the table.

Week	1	2	3	4	5	6
Amount saved	£1	£2	£3	£4	£5	£6
Total amount saved	£1	£3				

c Work out who will have more money at the end of 10 weeks.

5 This is a sequence of huts made from matches.

1 hut	2 huts	3 huts
7 matches	13 matches	19 matches

a Work out the rule for the nth term (the number of matches in the nth hut).

b Work out how many matches will be needed for the 30th hut.

c I use 91 matches to make a sequence of huts. How many huts do I make?

d I have 200 matches. What is the largest number of huts I can make? How many matches will be left over?

6 Work out the nth term for each sequence.

a 5, 14, 23, 32, 41, …, …   **b** 12, 10, 8, 6, 4, …, …   **c** 3.5, 4, 4.5, 5, 5.5, …, …

7 The patterns in this sequence are made from blue and yellow triangles.

Pattern 1   Pattern 2   Pattern 3   Pattern 4

a Copy and complete this table.

Pattern (term) number	Number of blue triangles	Number of yellow triangles	Total number of triangles
1	1	3	4
2	3		
3			
4			

b Describe the sequence formed by the numbers of blue triangles.

c Describe the sequence formed by the numbers of yellow triangles.

d Describe the sequence formed by the total numbers of triangles.

e Work out the total number of triangles in Pattern 10.

f What is the total number of triangles in the nth pattern?

Mathematical reasoning

Valencia Planetarium

This is the planetarium in the city of Valencia in Spain. Many of the features of the building are based on a repeating sequence.

This key explains the diagrams on these pages.

Ladders and grids are made from combinations of:

L links   T links   X links   R rods

Each combination can be expressed as a rule, using letters.

Example

This grid is a combination of L links, T links, X links and R rods.

The combination can be written as the rule:
$4L + 6T + 2X + 17R$

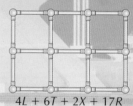

$4L + 6T + 2X + 17R$

1 Look at the ladders on the right.

  a Use letters to write down a rule for each of them.

  b Copy and complete this table.

Ladder	L links	T links	R rods
1	4	0	4
2	4	2	7
3			
4			
5			

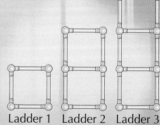

Ladder 1   Ladder 2   Ladder 3

  c Use letters to write down a rule for the links and rods in ladder 10.

  d Use letters to write down a rule for the links and rods in ladder n.

2 These rectangles are 2 squares deep.

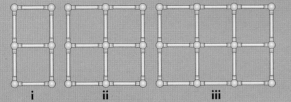

i   ii   iii

a Use letters to write down a rule for each of them.

b Copy and complete this table.

Rectangle	L links	T links	X links	R rods
2 by 1	4	2	0	7
2 by 2	4	4	1	12
2 by 3				
2 by 4				
2 by 5				

c Use letters to write down a rule for the links and rods in a 2 by 10 rectangle.

d Use letters to write down a rule for the links and rods in a 2 by n rectangle.

3 These rectangles are 3 squares deep.

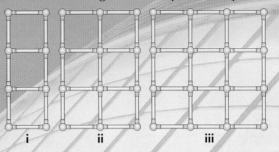

i   ii   iii

a Use letters to write down a rule for each of them.

b Copy and complete this table.

Rectangle	L links	T links	X links	R rods
3 by 1	4	4	0	10
3 by 2	4	6	2	17
3 by 3				
3 by 4				
3 by 5				

c Use letters to write down a rule for the links and rods in a 3 by 10 rectangle.

d Use letters to write down a rule for the links and rods in a 3 by n rectangle.

e Now write down a rule for the number of links and rods in a 4 by n rectangle.

4 a Use letters to write down a rule for each of these squares.

b Copy and complete this table.

Square	L links	T links	X links	R rods
1 by 1	4	0	0	4
2 by 2	4	4	1	12
3 by 3				

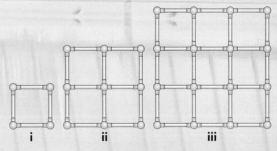

i   ii   iii

c Continue the table to work out the number of links and rods in a 7 by 7 square.

d Describe how the pattern of T links is building up.

e What type of special sequence do the numbers of X links make?

f Show that the number of rods in an n by n square is given by the nth term $2n(n + 1)$.

g Use letters to write down a rule for the number of links and rods in an 8 by 8 square.

h Use letters to write down a rule for the number of links and rods in an n by n square.

3

Perimeter, area and volume

This chapter is going to show you:

- how to work out the perimeters and areas of rectangles
- how to work out the perimeters and areas of compound shapes
- how to work out the areas of triangles, parallelograms and trapezia
- how to use simple formulae to work out perimeter, area and volume
- how to work out the surface areas, volumes and capacities of cuboids.

You should already know:

- that the perimeter of a shape is the distance around its edge
- that area is the space inside a flat shape
- the names of 3D shapes such as the cube and the cuboid.

About this chapter

'Pentagon' is the name for a five-sided shape. It is also the name of one of the biggest office buildings in the world, which is the headquarters of the USA Defence Department. The Pentagon has five sides and its perimeter is about 1.4 kilometres long. Its buildings cover an area of 600 000 m², with a central plaza of 20 000 m². Inside, the lengths of its corridors alone total 28.2 kilometres and over 30 000 people work there.

How does this compare to the area of your house and the perimeter of your bedroom?

3.1 Perimeter and area of rectangles

Learning objectives

- To use a simple formula to work out the perimeter of a rectangle
- To use a simple formula to work out the area of a rectangle

The **metric units** of **length** in common use are:

- the millimetre (mm)
- the centimetre (cm)
- the metre (m)
- the kilometre (km)

The metric units of **area** in common use are:

- the square millimetre (mm^2)
- the square centimetre (cm^2)
- the square metre (m^2).

A **rectangle** has two long sides and two short sides. The long sides are equal and each is referred to as the length of the rectangle. The short sides are also equal and each is referred to as the **width**.

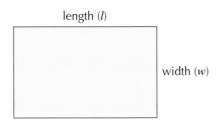

length (*l*)

width (*w*)

This means that its **perimeter** equals 2 × length + 2 × width.

Perimeter = 2 lengths + 2 widths

You can write this as a **formula**, using the letters *P* for perimeter, *l* for length and *w* for width.

$P = 2l + 2w$

The units are millimetres (mm), centimetres (cm) or metres (m).

You can find the area of a rectangle by multiplying its length by its width.

Area = length × width

You can write this as a formula, where *A* = area, *l* = length and *w* = width.

$A = l \times w$ or $A = lw$

The units are square millimetres (mm^2), square centimetres (cm^2) or square metres (m^2).

Using this formula, if you know the area of a shape you can work back from it to calculate its length or width if you do not already know it.

Example 1

Work out the perimeter (P) and area (A) of this wall tile.

$P = 2l + 2w$

$P = 2 \times 6 + 2 \times 4$

$\quad = 12 + 8$

$\quad = 20$ cm

$A = lw$

$\quad = 6 \times 4$

$\quad = 24$ cm^2

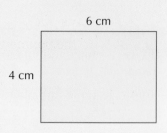

6 cm

4 cm

The length and width of a square are the same.

Example 2

Work out the perimeter and area of this square patio.

$P = 5 + 5 + 5 + 5$

$\quad = 4 \times 5$

$\quad = 20$ m

$A = 5 \times 5$

$\quad = 5^2$

$\quad = 25$ m^2

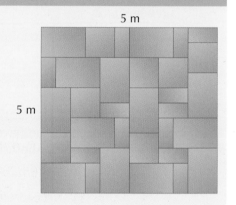

5 m

5 m

Example 3

This fence has an area of 21 m^2.

Work out the height of the fence, shown as h on the diagram.

The fence is a rectangle and its height is its shorter side, so take this as its width.

Area = length × width ($A = lw$)

So $21 = 7 \times w$

To find w, divide both sides of the equation by 7.

$\frac{21}{7} = w$

$3 = w$

So the height of the fence is 3 metres.

You could simply work out $21 \div 7 = 3$.

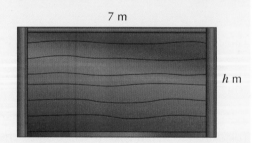

7 m

h m

1 Work out:

i the perimeter   ii the area of each rectangle.

a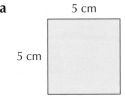
5 cm

5 cm

b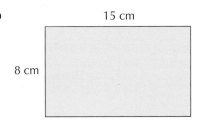
15 cm

8 cm

c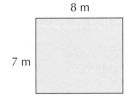
8 m

7 m

d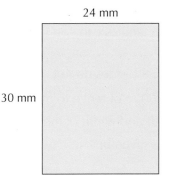
24 mm

30 mm

2 A football pitch measures 100 m by 75 m.

Work out the area of the pitch.

 3 A room measures 6 m by 4 m.

a What is the area of the floor?

b The floor is to be covered with square carpet tiles measuring 50 cm by 50 cm.

How many tiles are needed to cover the floor?

4 Work out the perimeter of this square tile.

25 cm²

5 Work out the length of each rectangle.

a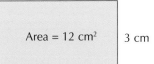
Area = 12 cm²   3 cm

b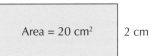
Area = 20 cm²   2 cm

c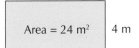
Area = 24 m²   4 m

d
Area = 48 cm²   6 cm

6 Copy and complete the table for rectangles **a** to **f**.

	Length	Width	Perimeter	Area
a	7 cm	5 cm		
b	30 cm	12 cm		
c	16 cm		50 cm	
d		$2\frac{1}{2}$ m	15 m	
e	7.5 m			45 m²
f		25 mm		800 mm²

(MR) 7 Draw diagrams to show that 1 cm² = 100 mm² and 1 m² = 10 000 cm².

(PS) 8 How many rectangles can you draw with a fixed perimeter of 20 cm, each having a different area?

The lengths of the sides do not need to be whole numbers.

(PS) 9 A farmer has 60 m of fence to make a rectangular sheep pen against a wall.

Find the length and width of the pen in order to make its area as large as possible.

An example is given below.

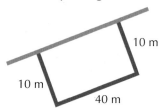

10 m

10 m

40 m

Investigation: Perimeter equals area?

Investigate whether a rectangle can have the same numerical value for its perimeter and its area.

3.2 Perimeter and area of compound shapes

Learning objective

- To work out the perimeter and the area of a compound shape

Key word

compound shape

A **compound shape** is made from more than one shape.

You can work out its perimeter and area by dividing it into the shapes that make it up.

Example 4

Work out the perimeter (*P*) and area (*A*) of this compound shape.

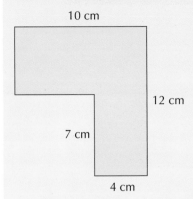

First split the shape into two rectangles. This split depends on the information you are given. If you split this shape into rectangles A and B, as shown, you will be able to work out all the lengths you need.

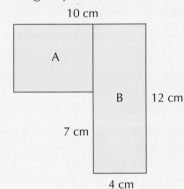

For rectangle A, the length is (10 − 4) = 6 cm and the width is (12 − 7) = 5 cm.

For rectangle B, the length is 12 cm and the width is 4 cm.

$P = 10 + 12 + 4 + 7 + 6 + 5 = 44$ cm

Total area = area of A + area of B

$$= 6 \times 5 + 12 \times 4$$
$$= 30 + 48$$
$$= 78 \text{ cm}^2$$

Example 5

Work out the area of the region shaded yellow in this shape.

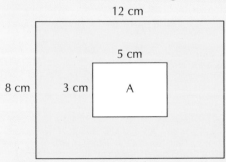

Area of complete rectangle = 12 × 8

                   = 96 cm²

Area of inner rectangle A = 5 × 3

                     = 15 cm²

So the area of the region shaded yellow = 96 − 15

                                         = 81 cm²

1 Work out:

i the perimeter   **ii** the area of each compound shape.

a

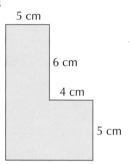

5 cm

6 cm

4 cm

5 cm

b

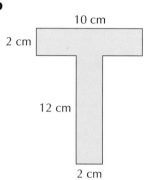

10 cm

2 cm

12 cm

2 cm

c

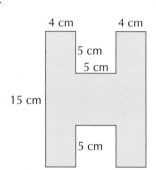

4 cm   4 cm

5 cm

5 cm

15 cm

5 cm

d

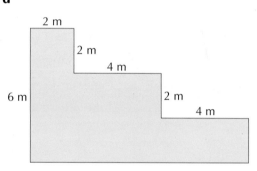

2 m

2 m

4 m

6 m

2 m

4 m

2 Work out the area of each compound shape.

a

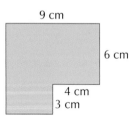

9 cm

6 cm

4 cm

3 cm

b

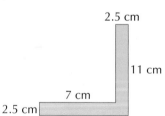

2.5 cm

11 cm

7 cm

2.5 cm

3 Danny works out the area of this compound shape.

This is his working.

Area = 12 × 5 + 10 × 6

    = 60 + 60

    = 120 cm²

 MR

a Explain why he is wrong.

b Work out the correct answer.

12 cm

5 cm

10 cm

6 cm

4 Kylie sticks a picture she took on her holiday in Paris onto a rectangular piece of card, like this.

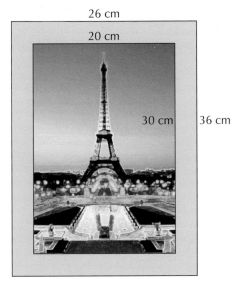

26 cm

20 cm

30 cm   36 cm

a Work out the area of the picture.

b Work out the area of the card she uses.

c Work out the area of the blue border.

(PS) 5 A garden is in the shape of a rectangle measuring 16 m by 12 m.

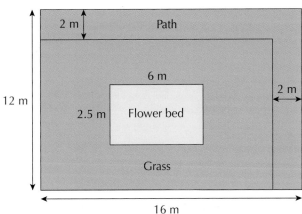

2 m   Path

6 m

2 m

12 m

2.5 m   Flower bed

Grass

16 m

Work out the area of the grass in the garden.

(PS) 6 Zoe is painting the kitchen wall in her house.

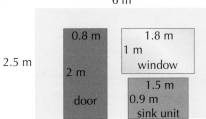

6 m

0.8 m   1.8 m

1 m

2.5 m

2 m   window

door   1.5 m

0.9 m

sink unit

Work out the area of the wall that needs to be painted.

Problem solving: Compound shapes

A This compound shape is made from two identical rectangles.

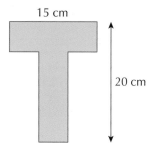

15 cm

20 cm

Work out the perimeter of the compound shape.

B The four unmarked sides in this diagram are the same length.

16 cm

16 cm

Work out the area of the compound shape.

3.3 Areas of some other 2D shapes

Learning objectives

- To work out the area of a triangle
- To work out the area of a parallelogram
- To work out the area of a trapezium

Key words	
base	height
parallelogram	perpendicular height
trapezium	triangle

Area of a triangle

To work out the area of a **triangle**, you need to know the length of its **base** and its **height**. You should measure the height by drawing a perpendicular line from the base to the angle above it. This means it is sometimes called the **perpendicular height**.

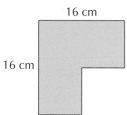

height (*h*)

base (*b*)

The diagram on the right shows that the area of the triangle is half of the area of a rectangle with the same base and height.

You can see that:

area 1 = area 2

area 3 = area 4

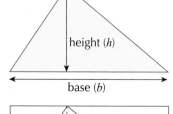

You can find the area of the rectangle by multiplying the base by the height.

58   3 Perimeter, area and volume

So, the area of the triangle is:

$\frac{1}{2} \times$ base $\times$ height

You can write this as a formula, where A represents the area, b is the base and h is the height:

$A = \frac{1}{2} \times b \times h$ or $A = \frac{1}{2}bh$

This is true for all triangles.

Example 6

Work out the area of this triangle.

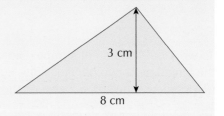

The formula for the area of a triangle is:

$A = \frac{1}{2}bh$

$A = \frac{1}{2} \times 8 \times 3$

$\quad = 4 \times 3$

$\quad = 12 \text{ cm}^2$

Example 7

Work out the area of this obtuse-angled triangle.

$A = \frac{1}{2}bh$

$A = \frac{1}{2} \times 10 \times 15$

$\quad = 5 \times 15$

$\quad = 75 \text{ cm}^2$

Hint Notice that you have to measure the perpendicular height outside the triangle.

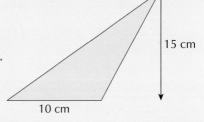

Area of a parallelogram

To work out the area of a **parallelogram**, you need to know the length of its base and its perpendicular height.

The diagrams below show that the parallelogram has the same area as a rectangle with the same base and height.

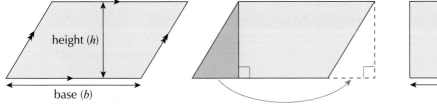
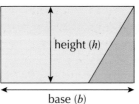

So, the area of a parallelogram is:

base $\times$ height

You can write this as a formula, where A represents the area, b is the base and h is the height:

$A = b \times h$ or $A = bh$

Example 8

Work out the area of this parallelogram.

The formula for the area of a parallelogram is:

$A = bh$

$A = 6 \times 10$

$\quad = 60 \text{ cm}^2$

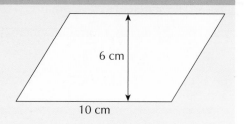

Area of a trapezium

To work out the area of a **trapezium**, you need to know the length of its two parallel sides, a and b, and the perpendicular height, h, between the parallel sides.

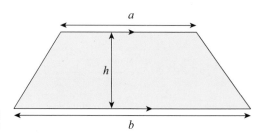

The diagram below shows how you can fit the trapezium together with another one that is exactly the same, to form a parallelogram. The area of each trapezium is half the area of the parallelogram.

You know that the area of a parallelogram is base times height.

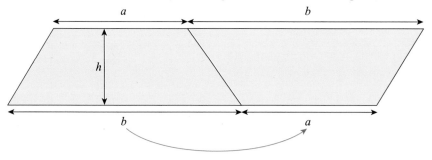

So, the area of a trapezium is:

$\frac{1}{2} \times$ sum of the lengths of the parallel sides $\times$ height

You can write this as a formula, where A represents the area, a and b are the lengths of the parallel sides and h is the height:

$A = \frac{1}{2} \times (a + b) \times h$ or $A = \frac{1}{2}(a + b)h$

 Hint You must add the numbers in the brackets before doing anything else.

Example 9

Work out the area of this trapezium.

The formula for the area of a trapezium is:

$A = \frac{1}{2}(a + b)h$

$A = \frac{1}{2} \times (9 + 5) \times 4$

$\quad = \dfrac{14 \times 4}{2}$

$\quad = 28 \text{ cm}^2$

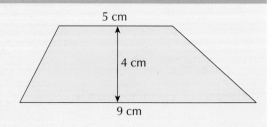

Exercise 3C

1 Work out the area of each triangle.

a

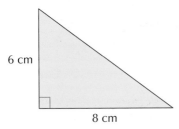

6 cm

8 cm

b

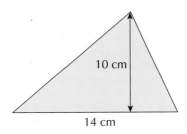

10 cm

14 cm

c

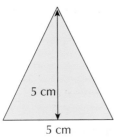

5 cm

5 cm

d

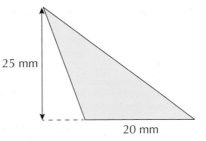

25 mm

20 mm

e

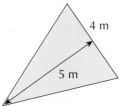

4 m

5 m

2 Copy and complete the table for triangles **a** to **e**.

	Base	Height	Area
a	5 cm	4 cm	
b	7 cm	2 cm	
c	9 m	5 m	
d	12 mm		60 mm²
e		8 m	28 m²

3 Work out the area of each parallelogram.

a

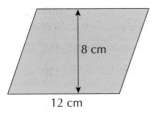

8 cm

12 cm

b

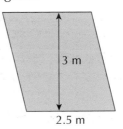

3 m

2.5 m

c

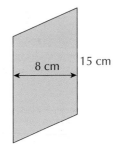

8 cm

15 cm

d

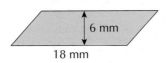

6 mm

18 mm

e

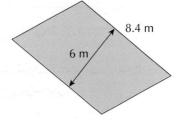

8.4 m

6 m

4 Work out:

i the perimeter   ii the area of this parallelogram.

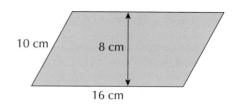

10 cm

8 cm

16 cm

5 Work out the area of each trapezium.

a

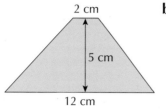

2 cm

5 cm

12 cm

b

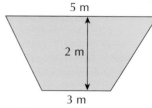

4 cm

8 cm

10 cm

c

5 m

2 m

3 m

d

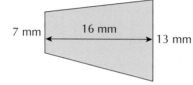

7 mm   16 mm   13 mm

e

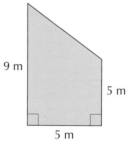

9 m

5 m

5 m

6 The diagram shows the end wall of a garden shed.

a Work out the area of the door.

b Work out the area of the brick wall.

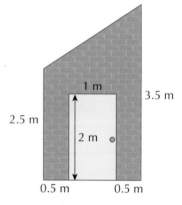

1 m

3.5 m

2.5 m

2 m

0.5 m   0.5 m

7 The diagram shows the measurements of a label on a sauce bottle.

Work out the area of the label.

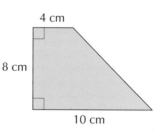

25 mm

52 mm

Tomato

SAUCE

20mm

30 mm

8 Work out the area of this computer worktop.

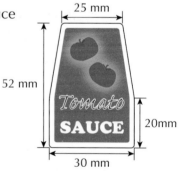

2.4 m

0.8 m

1.8 m

1 m

0.8 m

9 This mathematical stencil is made from a rectangle of plastic, with the shapes cut out.
Work out the area of plastic used.

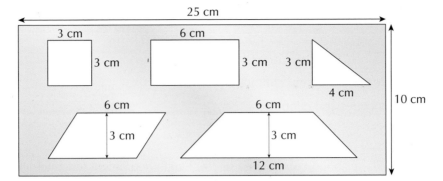

MR **10** The area of this trapezium is 8 cm².
Work out different values of a, b and h,
assuming that $b > a$.

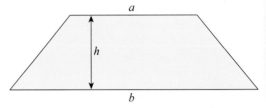

Investigation: Pick's formula

The shapes below are drawn on a centimetre-square grid of dots.

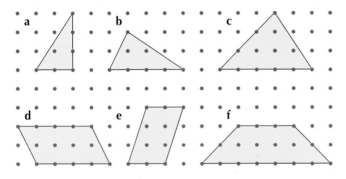

	Number of dots on perimeter of shape	Number of dots inside shape	Area of shape (cm²)
a			
b			
c			
d			
e			
f			

A Copy and complete the table for each shape.

B Work out a formula that connects the number of dots on the perimeter (P), the number of dots inside (I) and the area (A) of each shape.

C Check your formula by drawing different shapes on a centimetre-square grid of dots.

Georg Pick (1859–1942) was an Austrian mathematician. His simple formula allows us to find the area of any polygon drawn on a lattice or a grid.

3.4 Surface area and volume of cubes and cuboids

Learning objectives

- To work out the surface area of cubes and cuboids
- To work out the volume of cubes and cuboids

Key words

capacity	cube
cuboid	surface area
volume	

Shapes that are squares in 3D are called **cubes**. Their width, length and height (edge lengths) are all the same.

Shapes that are rectangles in 3D are called **cuboids**. Their width, length and height can all be different.

Surface area

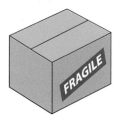 

To find the **surface area** of a cuboid you need to work out the total area of its six faces.

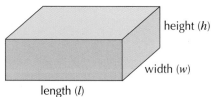

height (h)

width (w)

length (l)

Area of top and bottom faces = 2 × length (l) × width (w) = $2lw$

Area of front and back faces = 2 × length × height (h) = $2lh$

Area of the two sides = 2 × width × height = $2wh$

So the surface area S of a cuboid is:

$$S = 2lw + 2lh + 2wh$$

Example 10

Work out the surface area of this cuboid.

The formula for the surface area of a cuboid is:

$S = 2lw + 2lh + 2wh$

$= (2 \times 5 \times 4) + (2 \times 5 \times 3) + (2 \times 4 \times 3)$

$= 40 + 30 + 24$

$= 94 \text{ cm}^2$

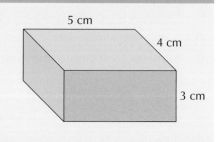

5 cm

4 cm

3 cm

Volume

Volume is the amount of space occupied by three-dimensional (3D) objects.
3D objects have height as well as length and width.

The diagram shows a cuboid that measures 4 cm by 3 cm by 2 cm.

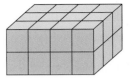

It is made up of cubes of edge length 1 cm. The top layer has 12 cubes. There are two layers so the cuboid has 24 cubes altogether. This number is the same as you would get by multiplying all its edge lengths: $4 \times 3 \times 2 = 24$.

So you can work out the volume of a cube or cuboid by multiplying its length by its width by its height.

Volume of a cube or cuboid = length × width × height

You can also write this as a formula:

$V = l \times w \times h = lwh$

where V = volume, l = length, w = width and h = height.

The metric units of volume in common use are:

- the cubic millimetre (mm^3)
- the cubic centimetre (cm^3)
- the cubic metre (m^3)

So, in the example above the volume of the cuboid is 24 cm^3. The cubes that make it up each have a volume of 1 cm^3.

The **capacity** of a 3D shape is the volume of liquid or gas it can hold. The metric unit of capacity is the litre (l). You need to know these metric conversions between capacity and volume.

- 1000 millilitres (ml) = 1 litre
- 1 litre = 1000 cm^3
- 1 ml = 1 cm^3
- 1000 l = 1 m^3

Example 11

Work out the volume of this tank and then work out its capacity, in litres.

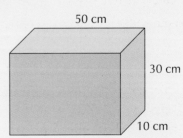

The formula for the volume of a cuboid is:

$V = lwh$

$V = 50 \times 30 \times 10$

$\quad = 15\,000 \text{ cm}^3$

$1000 \text{ cm}^3 = 1$ litre

So the capacity of the tank = $15\,000 \div 1000$ litres.

Capacity = 15 litres

Example 12

Work out the volume and capacity of this shape.

Give your answer in litres.

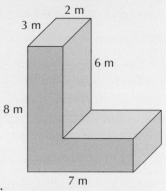

The shape is made up of two cuboids with measurements
7 m by 3 m by 2 m and 2 m by 3 m by 6 m.

$V = lwh$ so the volume of the shape is given by:

$V = (7 \times 3 \times 2) + (2 \times 3 \times 6)$

$\quad = 42 + 36$

$\quad = 78 \text{ m}^3$

As 1 m³ = 1000 litres, the capacity of the shape = 78 × 1000 litres.

Capacity = 78 000 litres

Exercise 3D

1 Work out: **i** the surface area **ii** the volume of each cuboid.

a
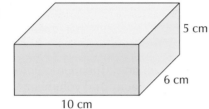

b
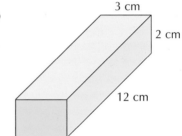

c
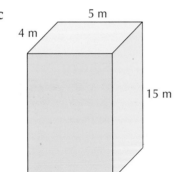

d
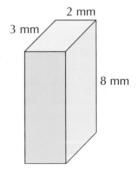

2 Work out:

i the surface area **ii** the volume of this unit cube.

3 Work out:

i the surface area **ii** the volume of the cubes with these edge lengths.

a 2 cm **b** 5 cm **c** 10 cm **d** 8 m

4 The diagram shows the dimensions of a swimming pool.

a Work out the volume of the pool, giving
the answer in cubic metres.

b How many litres of water does the pool
hold when it is full?

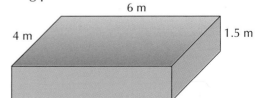

5 The diagram shows the dimensions of a rectangular carton of orange juice.

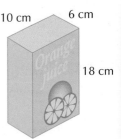

10 cm   6 cm

18 cm

  a Work out the volume of the carton, giving your answer in cubic centimetres.

  b How many glasses can be filled with orange juice from four full cartons, if each glass holds 240 ml?

6 How many packets of sweets, each measuring 8 cm by 5 cm by 2 cm, can be packed into a cardboard box measuring 32 cm by 20 cm by 12 cm?

 7 Twenty unit cubes are arranged to form a cuboid.

  a How many different cuboids can you make?

  b Which one has the greatest surface area?

8 Work out the surface area and volume of this 3D shape.

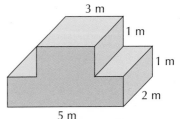

3 m

1 m

1 m

2 m

5 m

 9 The diagram shows the areas of the faces of a cuboid.

Use this information to work out the volume of the cuboid.

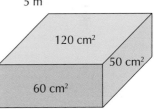

120 cm²

50 cm²

60 cm²

Investigation: Volume of an open box

An open box is made from a piece of card, measuring 20 cm by 16 cm, by cutting off identical squares from each corner.

Investigate the volume of the open box formed when you cut off different sizes of square.

You could put your data in a table, like the one below, or use a computer spreadsheet.

Comment on your results.

20 cm

16 cm

Size of square	Length of box	Width of box	Height of box	Volume of box
1 cm by 1 cm	18 cm	14 cm	1 cm	252 cm³
2 cm by 2 cm	16 cm	12 cm	2 cm	384 cm³

Ready to progress?

Review questions

1   The diagram shows square A. Fold square A in half to make rectangle B.
    Then fold rectangle B in half to make square C.

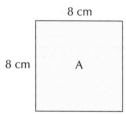

8 cm

8 cm   A

B

C

Copy and complete the table below, to show the area and perimeter of each shape.

	Area	Perimeter
Square A	... cm²	... cm
Rectangle B	... cm²	... cm
Square C	... cm²	... cm

2   A rectangle has an area of 48 cm².

    The length is three times the width.

    Work out the length of the rectangle.

3   A shop sells square carpet tiles, as shown opposite.

    The floor of a rectangular room is 3 m by 1.5 m.

    How many tiles are needed to carpet the floor?

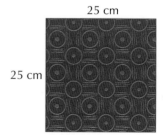

25 cm

25 cm

4   Work out the area of each shape.

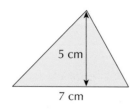

5 cm

7 cm

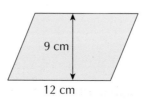

9 cm

12 cm

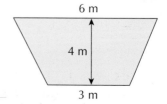

6 m

4 m

3 m

5   The diagram shows a square, cut into four rectangles.
     The area of one of the rectangles is shown.
     Work out the area of the rectangle marked A.

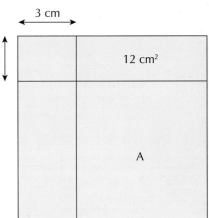

3 cm

2 cm

12 cm²

A

PS 6   This shape is made from four identical rectangles.

12 cm

     Work out the area of the shape.

PS 7   Work out the area of each compound shape below.

a

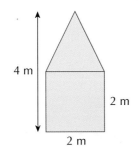

4 m

2 m

2 m

b

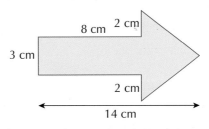

8 cm   2 cm

3 cm

2 cm

14 cm

c

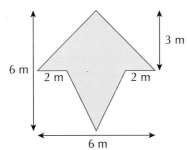

3 m

6 m   2 m   2 m

6 m

8   The two cuboids A and B have the same volume.

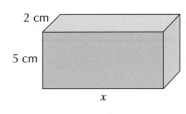

Cuboid A

5 cm

3 cm

4 cm

Cuboid B

2 cm

5 cm

x

a   Work out: i the surface area   ii the volume of Cuboid A.
b   Work out the value of the length marked x on Cuboid B.

Problem solving
Design a bedroom

1 This is a sketch plan for a bedroom.

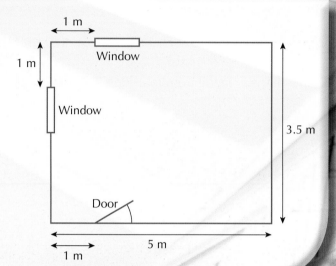

    a What is the perimeter of the bedroom?

    b What is the area of the bedroom?

    c One square metre of carpet costs £32.50. How much does it cost to carpet the bedroom?

2 Posters cost £6.99 each.

    a How many posters can you buy for £50?

    b How much money will you have left over?

3 These are sketches of the door and one of the windows.

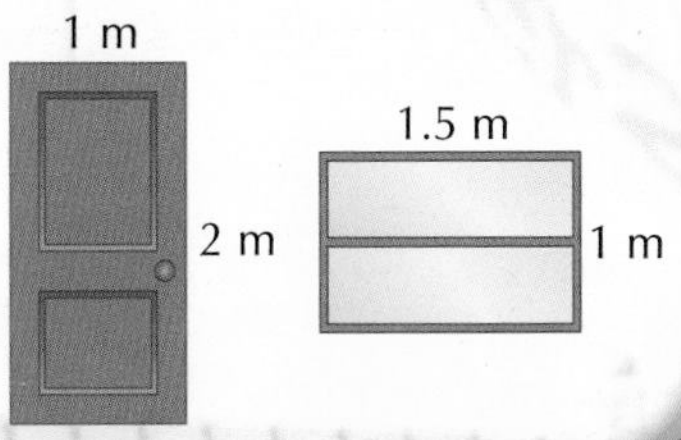

The height of the bedroom is $2\frac{1}{2}$ m.

1 m

1.5 m

2 m   1 m

a What is the total area of all four walls?

b A one-litre tin of paint covers 12 m². What is the minimum number of tins required to paint the walls?

4 Ellie wants to buy a storage unit for her collection of DVDs, for her bedroom.

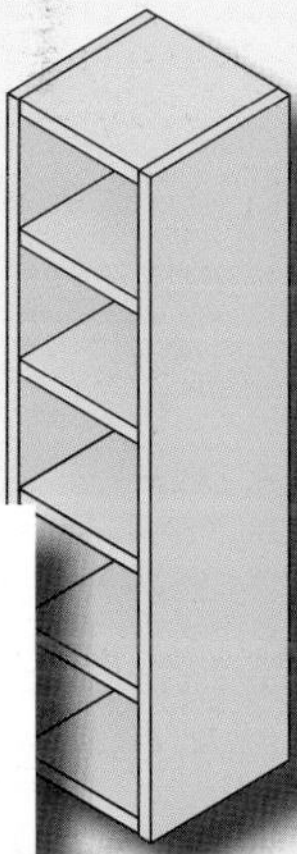

a This unit is 1 m high, 21 cm wide and has a depth of 17 cm.

Work out the volume of the unit.

b Ellie wants to store her collection of DVDs in this unit.

The thickness of the wood in the unit is 1.5 cm.

The thickness of each DVD is 1.5 cm.

What is the maximum number of DVDs she can store?

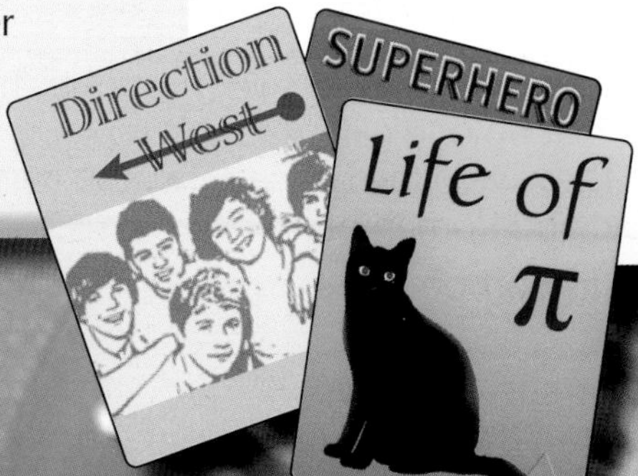

Direction West

SUPERHERO

Life of π

4

Decimal numbers

This chapter is going to show you:

- how to order numbers with up to four places of decimals, by size
- how to multiply and divide decimal numbers by 10, 100, 1000 and 10 000
- how to use estimation to check your answers
- how to round a decimal number, correct to one decimal place
- how to add and subtract decimal numbers
- how to multiply and divide decimals by decimals.

You should already know:

- how to write and read whole numbers and decimals
- how to write tenths and hundredths as decimals
- multiplication tables up to 12×12
- how to use a calculator to do simple calculations
- how to do short multiplication and division.

About this chapter

The decimal number system is based on 10. Decimals have been used for so long that no one can say exactly where or when they started. They were the basis of the ancient Chinese, Hindu-Arabic and Roman number systems.

The numbers we use in Europe come from the Hindu-Arabic number system.

Often we refer to decimals when we mean decimal fractions: tenths, hundredths, thousandths and so on.

We express these as 0.1, 0.01, 0.001, etc.

4.1 Multiplying and dividing by 10, 100, 1000 and 10 000

Learning objective

- To multiply and divide decimal numbers by 10, 100, 1000 and 10 000

Key words

| decimal | decimal point |

You already know how to multiply whole numbers by ten, a hundred, a thousand, ten thousand and so on. You can use the same method to multiply **decimals**.

When you multiply by 10, all the digits move one place to the left.

For example: $4 \times 10 = 40$ $4.15 \times 10 = 41.5$

When you multiply by 100, all the digits move two places to the left.

For example: $4 \times 100 = 400$ $4.15 \times 100 = 415$

When you multiply by 1000, all the digits move three places to the left.

For example: $4 \times 1000 = 4000$ $4.15 \times 1000 = 4150$

When you multiply by 10 000, all the digits move four places to the left.

For example: $4 \times 10\ 000 = 40\ 000$ $4.15 \times 10\ 000 = 41\ 500$

Notice that the number of places the digits move left is the same as the number of zeros in the number you are multiplying by. This is true when multiplying decimals as well as whole numbers.

The **decimal point** separates the whole-number part of the number from the tenths, hundredths and thousandths. The decimal place after thousandths is ten thousandths, then hundred thousandths and so on.

Example 1

Work these out.

a 3.5×100 **b** 4.7×10

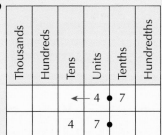

a

Thousands	Hundreds	Tens	Units	Tenths	Hundredths
	←	—	3 •	5	
	3	5	0 •		

The digits move two places to the left when you multiply by 100.

$3.5 \times 100 = 350$

b

Thousands	Hundreds	Tens	Units	Tenths	Hundredths
		←	4 •	7	
		4	7 •		

The digits move one place to the left when you multiply by 10.

$4.7 \times 10 = 47$

You can divide decimals by 10, 100, 1000 and 10 000 in the same way as you divide whole numbers.

When you divide by 10, all the digits move one place to the right.

For example: $37 \div 10 = 3.7$ $621.8 \div 10 = 62.18$

When you divide by 100, all the digits move two places to the right.

For example: $37 \div 100 = 0.37$ $621.8 \div 100 = 6.218$

When you divide by 1000, all the digits move three places to the right.

For example: $37 \div 1000 = 0.037$ $621.8 \div 1000 = 0.6218$

When you divide by 10 000, all the digits move four places to the right.

For example: $37 \div 10\,000 = 0.0037$ $621.8 \div 10\,000 = 0.062\,18$

Notice that the number of places the digits move right is the same as the number of zeros in the number you are dividing by.

Example 2

Work these out.

a $23 \div 100$ **b** $13.6 \div 1000$

a

Tens	Units	Tenths	Hundredths
2	3		
	0	2	3

b

Tens	Units	Tenths	Hundredths	Thousandths	Ten-thousandths
1	3	6			
	0	0	1	3	6

The digits move two places to the right when you divide by 100.

$23 \div 100 = 0.23$

The digits move three places to the right when you divide by 1000.

$13.6 \div 1000 = 0.0136$

Exercise 4A

1 Work these out.

a 3.4×10 **b** 0.045×10 **c** 0.6×10 **d** 0.89×100

e 0.053×100 **f** 0.03×100 **g** $0.4 \div 1000$ **h** $5.8 \div 1000$

i $3.4 \div 10$ **j** $0.045 \div 10$ **k** $0.6 \div 10$ **l** $0.89 \div 100$

2 Copy each statement, filling in the missing operation in the box and the missing number in the space.

a $0.37 \,\square\, \ldots = 370$ **b** $567 \,\square\, \ldots = 0.567$

c $0.07 \,\square\, \ldots = 70$ **d** $650 \,\square\, \ldots = 6.5$

3 Write down the missing number in each case.

a $0.07 \times 10 = \boxed{}$ **b** $0.7 \times \boxed{} = 70$ **c** $0.7 \div 10 = \boxed{}$

d $7 \div \boxed{} = 0.07$ **e** $0.7 \times 10 = \boxed{}$ **f** $0.07 \times \boxed{} = 700$

g $0.07 \div 100 = \boxed{}$ **h** $0.7 \div \boxed{} = 0.007$ **i** $\boxed{} \div 100 = 0.07$

4 To change from metres to centimetres, you multiply by 100.

For example, 4.7 m = 4.7 × 100 cm = 470 cm.

Write each length in centimetres.

a 7.8 m   **b** 4.55 m   **c** 29.5 m   **d** 0.8 m

e 0.14 m   **f** 2.085 m   **g** 0.0076 m   **h** 0.004 m

5 To change from grams to kilograms, you divide by 1000.

For example, 250 g = 250 ÷ 1000 kg = 0.25 kg.

Write each mass in kilograms.

a 1395 g   **b** 475 g   **c** 50 g   **d** 6270 g

e 81.5 g   **f** 20.08 g   **g** 125.5 g   **h** 70.009 g

6 10^2 means $10 \times 10 = 100$ and 10^3 means $10 \times 10 \times 10 = 1000$.

Copy and complete each statement.

$10^4 = \ldots \times \ldots \times \ldots \times \ldots = \ldots$

$10^5 = \ldots = \ldots$

$10^6 = = \ldots$

7 Write down the answers.

a 3.5×10^2 **b** 0.4×10^3 **c** 0.07×10^2 **d** 2.7×10^5

e 0.6×10 **f** 7.08×10^3 **g** $3.5 \div 10^2$ **h** $0.4 \div 10^3$

8 In 2012 the population of some countries were recorded, as shown in this table.

Country	Population
Spain	46.17 million
UK	62.744 million
Mexico	114.8 million
USA	311.59 million
China	1.344 billion

Copy the table, writing the numbers in full, for example, the population of Spain is 46 170 000.

9 In 2013 the government reported that the UK debt to banks around the world was 120.6 billion pounds.

One newspaper reported this as £1 206 000 000, which was incorrect. How should it have been reported?

4.2 Ordering decimals

Learning objective

- To order decimal numbers according to size

Key words	
order	place value
reciprocal	

Name	Len	Mia	Shehab	Baby Joe	Harry	Connie
Height	170 cm	1.58 m	189 cm	0.55 m	150 cm	1.80 m
Mass	75 kg	50.3 kg	68 kg	7.5 kg	75 kg	76 kg 300 g

Look at the people in the picture. How would you put their heights or masses in **order** of size?

Before you can compare them you need to rewrite each set of measures in the same units.

For example, the heights are given in metres and in centimetres. There are 100 centimetres in a metre so you can rewrite the heights given in centimetres as heights in metres by dividing them by 100.

Name	Len	Mia	Shehab	Baby Joe	Harry	Connie
Height	1.70 m	1.58 m	1.89 m	0.55 m	1.5 m	1.80 m

Now you can compare the size of the numbers. To do this you have to consider the **place value** of each digit.

Example 3

Put the numbers 2.33, 2.03, 2.3106, 2.0315 and 2.304 in order, from smallest to largest.

It helps to put the numbers in a table like this one.

Use zeros to fill the missing decimal places.

Working across the table from the left, you can see that all of the numbers have the same units digit. Three of them have the same tenths digit, three have the same hundredths digit and two have the same thousandths digit.

The smallest two are 2.03 and 2.0315. You can see this because both have 0 tenths and both have 3 hundredths, but 2.03 doesn't have any more digits after hundredths.

Therefore 2.03 is the smallest and 2.0315 is the next smallest.

Next is 2.304 because it has fewer hundredths than 2.33 and 2.3016, even though they all have the same number of tenths.

Thousands	Hundreds	Tens	Units	Tenths	Hundredths	Thousandths	Ten thousandths
			2	3	3	0	0
			2	0	3	0	0
			2	3	0	1	6
			2	0	3	1	5
			2	3	0	4	0

The last two are 2.33 and 2.3016. They both have the same number of tenths, but 2.3016 has fewer hundredths, so 2.33 is the largest.

In order, from smallest to largest, the numbers are: 2.03, 2.0315, 2.304, 2.3016, 2.33.

Example 4

Put the correct sign, > or <, between the numbers in each pair.

a 6.05 and 6.046 **b** 0.06 and 0.065 **c** 0.008 02 and 0.0081

a Both numbers have the same units and tenths digits, but the value of the hundredths digit is bigger in the first number.

So the answer is 6.05 > 6.046.

b Both numbers have the same units, tenths and hundredths digits, but the second number has the bigger thousandths value, as the first number has a zero in the thousandths.

So the answer is 0.06 < 0.065.

c Both have the same units, tenths, hundredths and thousandths digits, but the second number has a bigger ten-thousandths value.

So the answer is 0.008 02 < 0.0081.

Exercise 4B

1 Put the correct sign, > or <, between the numbers in each pair.

  a 0.416 … 0.42 **b** 0.72 … 0.702 **c** 5.79 … 5.709
  d 3.75 km … 3.225 km **e** 0.844 kg … 0.8 kg **f** £0.08 … 12p

2 Write these numbers in order, from smallest to largest.

  4.57, 0.0045, 4.057, 4.5, 0.0457, 0.5, 4.05

3 Write each set of numbers in order, from smallest to largest.

  a 0.0073, 0.073, 0.008, 0.7098, 0.7
  b 1.2033, 1.0334, 1.405, 1.4045, 1.4
  c 34, 3.4, 0.34, 0.034, 3.0034

4 In 2012:

- the estimated population of China was 1.35 billion
- the estimated population of India was 1 330 044 544.

Which country had the greater population?

5 Ben did some internet research and found this information about the average distance of each planet from the Sun, in our solar system.

Planet	Average distance from the Sun (km)
Earth	149.6 million
Jupiter	778 500 000
Mars	0.228 billion
Mercury	58 million
Neptune	4 500 000 000
Saturn	1.43 billion
Uranus	2 880 000 000
Venus	108 million

Copy the table, listing the planets in order of distance from the Sun and starting with the one closest to it.

6 One metre is 100 centimetres.

Put these lengths in order, from smallest to largest.

6.5 m, 269.8 cm, 32.68 cm, 27 m, 6485 cm

7 One kilogram is 1000 grams.

Put these masses in order, from smallest to largest.

4671 g, 4.7 kg, 999 g, 5 kg, 5500 g

8 One litre is 1000 millilitres.

Put these capacities in order, from smallest to largest.

7 litres, 650 millilitres, 6.85 litres, 7308 millilitres, 7.085 litres

Investigation: Reciprocals

The **reciprocal** of a number is the result of dividing it into 1. For example, these are the reciprocals of 5, 6, 7, 8 and 9.

$1 \div 5, 1 \div 6, 1 \div 7, 1 \div 8, 1 \div 9$

A   Use your calculator to work out the reciprocals of 5, 6, 7, 8 and 9.

B   Put the answers in order, from smallest to largest.

C   Repeat with five consecutive two-digit whole numbers, for example, 15, 16, 17, 18, 19.

D   What do you notice?

4.3 Estimates

Learning objectives

- To estimate calculations in order to spot possible errors
- To round up or down, to one decimal place

TOWN v CITY

CROWD	52 812
SCORE	3 – 2
TIME OF FIRST GOAL	18 min 27 sec
PRICE OF A PIE	£3.49
CHILDREN	33% off normal ticket prices

Suppose you were telling a friend about the game. Which of the numbers above would you round up or down to a sensible **approximation**? Which ones must you give exactly?

There are times when you want to **round** a number to the nearest thousand, as you might for the crowd size. 52 812 is closer to 53 000 than to 52 000 so you would round it up to that.

You can round to hundreds, tens or just to the nearest whole number, as you might for the time of the first goal. As 27 seconds is less than half a minute, the first goal was scored nearer to 18 minutes than 19 minutes into the game, so you would round it down to 18 minutes.

Sometimes you want to round decimal numbers. For example, a pie costs £3.49 but it is simpler to think of this rounded to £3.50. When it is not a price, you reduce the number of decimal places as well. So 3.49 would round to 3.5 and 3.42 to 3.4. This is rounding to one decimal place (1dp).

When a number is equally close to both the lower and the higher values, always round up. For example, 3.45 would round to 3.5.

You can also round numbers (up or down) to **estimate** quickly whether the answer to a calculation is about right. There are some other quick checks you can use.

- First, for a multiplication, you can check that the final digit is correct.
- Second, you can round numbers and do a mental calculation to see if an answer is about the right size.
- Third, you use the **inverse operation** (see Example 7).

Example 5

Explain why these calculations must be wrong.

a $23 \times 45 = 1053$ **b** $19 \times 59 = 121$

a The last digit should be 5, because the product of the last digits (3 and 5) is 15.

That is, $23 \times 45 = ...5$, so the calculation is wrong.

b The actual answer is roughly $20 \times 60 = 1200$, so the calculation is wrong.

Example 6

Estimate answers to these calculations.

a $\dfrac{21.3 + 48.7}{6.6}$ b 31.2×48.5 c $359 \div 42$

  a Round the numbers on the top, $20 + 50 = 70$. Round 6.6 to 7. Then $70 \div 7 = 10$.

  b Round to 30×50, which is $3 \times 5 \times 100 = 1500$.

  c Round to $360 \div 40$, which is $36 \div 4 = 9$.

Example 7

Use the inverse operation to check each calculation.

a $450 \div 6 = 75$ b $310 - 59 = 249$

  a By the inverse operation, $450 = 6 \times 75$.

    Check mentally.

    $6 \times 70 = 420$, $6 \times 5 = 30$, $420 + 30 = 450$, so is true.

  a By the inverse operation, $310 = 249 + 59$.

    This must end in 8 as $9 + 9 = 18$, so the calculation cannot be correct.

Example 8

Round each number to one decimal place.

a 8.746   b 3.172   c 6.35

To round to one decimal place (to the nearest tenth), look at the digit in the second decimal place (the hundredths).

Is it worth 5 or more? If so round the tenths up. If not, leave the tenths digit as it is.

  a 8.746　　The hundredths digit (second decimal place) is 4.

  This is less than 5, so leave the tenths digit as it is.

  $8.746 \approx 8.7$ (1dp)

  b 3.172　　The hundredths digit is 7.

  This is higher than 5, so round the tenths digit up.

  $3.172 \approx 3.2$ (1dp)

  c 6.35　　The hundredths digit is 5, so round up.

  $6.35 \approx 6.4$ (1dp)

Exercise 4C

1 Explain why each calculation must be wrong.

   a $32 \times 54 = 1782$ **b** $83 \times 41 = 323$ **c** $\dfrac{43.4 + 53.1}{9.6} = 20.05$

   d $52.5 \div 7 = 5.7$ **e** $453 - 87 = 393$ **f** $68.3 \times 108 = 776.4$

(PS) 2 A merchant bought 183 kg of grain at a cost of \$3.84 per kilogram.

   What is the approximate total cost of this grain?

3 Estimate the answer to each problem.

   a $1657 - 728$ **b** 342×29 **c** $681 \div 29$ **d** $\dfrac{47.8 + 73.3}{13.2}$

   e 312×534 **f** $268.3 \div 27.12$ **g** $\dfrac{246.8 - 79.3}{26.9 - 7.8}$ **h** $\dfrac{27.8 \times 72.1}{28.5 - 19.5}$

(MR) 4 Kristy bought eight cans of lemonade at 87p per can. The shopkeeper asked her for £7.26.

   Without working out the correct answer, explain how Kristy can see why this is wrong.

(MR) 5 A cake costs £1.37. I need six cakes.

   Will £10 be enough to pay for them? Explain your answer clearly.

(MR) 6 In a shop I bought a 55p stamp and a £3.45 birthday card. The total on the till was £58.45. Why?

7 Explain which of these is the best estimate for $64.5 \div 21.8$.

   a $50 \div 20$ **b** $64 \div 21$ **c** $60 \div 20$ **d** $63 \div 21$

8 Round each number to one decimal place.

   a   0.56   **b**   0.67   **c**   0.89   **d**   1.23   **e**   3.45

   f   1.38   **g**   4.72   **h**   9.99   **i**   0.12   **j**   0.07

   k   1.46   **l**   5.216   **m**   8.765   **n**   5.032   **o**   5.067

9 You can estimate $62 \div 0.39$ as $60 \div 0.4 = 600 \div 4 = 150$.

   Estimate the answer to each calculation.

   a $62 \div 0.56$ **b** $139 \div 0.67$ **c** $39 \div 0.81$ **d** $42 \div 0.17$

   e $57 \div 0.33$ **f** $67 \div 0.69$ **g** $38 \div 0.18$ **h** $178 \div 0.91$

   i $269 \div 0.86$ **j** $38 \div 0.75$ **k** $34 \div 0.52$ **l** $116 \div 0.18$

(PS) 10 Jess had £20. In her shopping basket she had a bag costing £6.85, some pens costing £5.85 and a book costing £5.99.

   a   Without adding up the numbers, how could Jess be sure she had enough money to buy the goods in the basket?

   b   Explain a quick way for Jess to find out if she could afford a 55p bar of chocolate as well.

11 These are the amounts of money a family of four brought home, as wages.

David $180 per week	Olivia $685 per month
James $290 per week	Kathryn $980 per month

To qualify for an educational bursary, they had to be bringing home a total of less than $50 000 per year. Use estimates to decide whether they will qualify.

Challenge

The first 15 square numbers are 1, 4, 9, 16, 25, 36, 49, 64, 81, 100, 121, 144, 169, 196 and 225.

The inverse operation of squaring a number is to find its **square root**. The sign for a square root is $\sqrt{}$. So $\sqrt{121} = 11$.

Only the square numbers have square roots that are whole numbers.

You have to find the square roots of other numbers by estimating or using a calculator.

For example, use a diagram like this one to estimate that $\sqrt{30} \approx 5.48$.

$$
\begin{array}{c}
x \quad 5 \quad\quad\quad 6 \\
\hline
x^2 \quad 25 \quad\quad 36 \\
\quad\quad 30
\end{array}
$$

A check shows that $5.48^2 = 30.03$ so this a good estimate.

Here is another example.

This diagram shows that $\sqrt{45} \approx 6.7$.

$$
\begin{array}{c}
x \quad 6 \quad\quad\quad 7 \\
\hline
x^2 \quad 36 \quad\quad 49 \\
\quad\quad 45
\end{array}
$$

A check shows that $6.7^2 = 44.89$ so this is a good estimate.

A Use the above method to estimate $\sqrt{19}$, $\sqrt{65}$, $\sqrt{95}$, $\sqrt{180}$, $\sqrt{240}$, $\sqrt{500}$.

B Use a calculator to check your answers.

4.4 Adding and subtracting decimals

Learning objective

• To add and subtract with decimal numbers

Key word

trailing zeros

You can add and subtract decimals in the same way as you do whole numbers. As with whole numbers, it is important to get the place values right.

If you are including whole numbers in additions and subtractions with decimals, it is helpful to write a decimal point after the units digit and show the decimal places with zeros. These are **trailing zeros**.

Example 9

Work these out.

a 18 + 5.86 + 0.978   **b** 6 − 1.45

  a 18 + 5.86 + 0.978

    The largest number of decimal places in the three numbers to be added is three, so add trailing zeros to the first two numbers.

    Then line up the decimal points and place values of all the numbers in the addition.

$$
\begin{array}{r}
18.000 \\
5.860 \\
+\,0.978 \\
\hline
24.838
\end{array}
$$

  b 6 − 1.45

    As in part **a**, add trailing zeros to the whole number to show the place values.

    Then line up the decimal points.

$$
\begin{array}{r}
{}^{5}\,{}^{9}\,{}^{1} \\
6.\!\!\not{0}\,\,0 \\
-\,1.4\,5 \\
\hline
4.5\,5
\end{array}
$$

Exercise 4D

1   Work these out.

    a   37.1 + 14.2   **b**   32.6 + 15.73   **c**   6.78 + 4.593   **d**   9.62 + 0.7

    e   4.79 + 1.2   **f**   6.084 + 2.16   **g**   1.2 + 3.41 + 4.563   **h**   76.57 + 312.5 + 6.089

2   Work out these subtractions.

    a   37.1 − 14.2   **b**   32.6 − 15.73   **c**   6.78 − 4.59   **d**   9.62 − 0.7

    e   4.79 − 1.2   **f**   6.08 − 2.16   **g**   1.2 + 3.41 − 4.56   **h**   76.57 + 312.5 − 6.08

3   Mr Li took his two children to the circus. The tickets cost £8.45 for an adult, £6.50 for the elder child and £4.15 for the younger.

    How much change did Mr Li get when he paid with a £20 note?

4   Work out the change you would get from £10 if you bought goods worth:

    a   £4.56   **b**   £3.99   **c**   £7.01   **d**   34p.

5 Tara checks her restaurant bill. She thinks it is wrong.

2 glasses of wine	£12.90
Soup of the day	£4.95
Chicken salad	£11.95
Beefburger	£10.85
Chargrilled steak	£14.55
Chips	£3.85
Grilled mushrooms	£3.65
Potato wedges	£3.45
Garlic ciabatta	£2.85
3 desserts	£19.15
Total	£95.15

a Use estimation to check if the bill is about right.

b Calculate if the bill is actually correct.

6 Estimate which of these has the greatest value, then use a calculator see if you were right.

a $25.41 + 9.75$ b $41 - 6.93$ c $\dfrac{136.34 + 246.57}{11}$

Challenge: Decimals in your head

Ahmed needed to work out $5 - 1.368$ without using paper and pencil or a calculator.

He had to do it in his head and tell someone else the answer. This is how he did it.

He first thought of 5 as 5.000, then worked out what must be added to 1.368 to get 5.000.

He started with the last decimal place and worked towards the left.

He thought: To get from 8 to 0, I must add 2, so the digit in the third place is 2.   _ . _ _ 2

I must carry 1, so add 1 to the digit in the second decimal place.

To get from 7 to 0, I must add 3, so the digit in the second place is 3.   _ . _ 32

I must carry 1, so add 1 to the digit in the first decimal place.

To get from 4 to 0, I must add 6, so the digit in the third place is 6.   _ . 632

I must carry 1, so add 1 to the units digit.

To get from 2 to 5, I must add 3, so the units digit is 3.   3.632

Ahmed did all that in his head and could just say that the answer was 3.632.

A Use Ahmed's method to work out each subtraction in your head, then simply write down the answer.

a $5 - 0.546$ b $8 - 2.672$ c $23 - 5.5643$

B When you have finished, use a calculator to check your answers and see how accurate you were.

4.5 Multiplying decimals

Learning objective

• To multiply decimal numbers

When you multiply a decimal by a whole number you do it like any other multiplication.

When you multiply two decimals together, the easiest method is to:

• count up all the decimal places in the numbers you are multiplying together
• do the multiplication as if they were whole numbers
• write the result with the same total number of decimal places as you started with, in the numbers you multiplied.

Example 10

Work these out.

a 3.7×9 **b** 6.24×8

As this is a decimal number multiplied by a whole number, you can do these as 'short' multiplications and keep the decimal points lined up.

$$
\begin{array}{r}
\textbf{a} \quad 3.7 \\
\times 9 \\
\hline
33.3 \\
{\scriptstyle 6}
\end{array}
\qquad
\begin{array}{r}
\textbf{b} \quad 6.24 \\
\times \ 8 \\
\hline
49.92 \\
{\scriptstyle 1 \ 3}
\end{array}
$$

Example 11

Work these out.

a 0.8×0.7 **b** 0.02×0.7

a As this is a decimal multiplied by a decimal, count the decimal places in both numbers (there are two), then calculate the product without the decimal points.

Then put the two decimal places back into the answer.

$8 \times 7 = 56$, so $0.8 \times 0.7 = 0.56$

This is because the number of decimal places in the answer is always the same as in the original calculation.

b $2 \times 7 = 14$, and there are three decimal places in the original calculation.
$0.02 \times 0.7 = 0.014$

Notice that a zero has been put in to make up the correct number of decimal places.

Example 12

Work these out.

a 2000×0.07 **b** $0.06 \times 50\,000$

    a $2000 \times 0.07 = 200 \times 0.7 = 20 \times 7 = 140$

    If you divide one number by 10 and multiply the other by 10, then their product will remain the same.

    b $0.06 \times 50\,000 = 0.6 \times 5000 = 6 \times 500$

    These are all equivalent calculations, so the answer is $6 \times 500 = 3000$.

Exercise 4E

1   Write down the answer to each multiplication.

    a 0.2×0.3 **b** 0.4×0.2 **c** 0.6×0.6 **d** 0.7×0.2

    e 0.02×0.4 **f** 0.8×0.04 **g** 0.06×0.1 **h** 0.3×0.03

    i 0.7×0.8 **j** 0.07×0.08 **k** 0.9×0.3 **l** 0.006×0.9

    m 0.5×0.09 **n** 0.5×0.5 **o** 0.8×0.005 **p** 0.06×0.03

2   Work out each multiplication.

    a 300×0.8 **b** 0.06×200 **c** 0.6×500 **d** 0.02×600

    e 0.03×400 **f** 0.004×500 **g** 0.007×200 **h** 0.002×9000

    i 0.005×8000 **j** 200×0.006 **k** 300×0.01 **l** 800×0.06

    m 500×0.5 **n** 400×0.05 **o** 300×0.005 **p** 200×0.0005

3   Work out each multiplication.

    a $0.006 \times 400 \times 200$ **b** $0.04 \times 0.06 \times 50\,000$ **c** $0.2 \times 0.04 \times 300$

    d $300 \times 200 \times 0.08$ **e** $20 \times 0.008 \times 40$ **f** $0.1 \times 0.07 \times 2000$

4   Work out the cost of six thousand CDs that are priced at £2.95 each.

5   A bottle of grape juice costs £2.62.

    Work out the cost of nine hundred bottles.

6   Bolts cost £0.06 each. An engineering company orders 20 000 bolts.

    How much will this cost?

7   A grain of sand weighs 0.006 grams.

    How much would 500 000 grains weigh?

8   A kilogram of uranium ore contains 0.000 002 kg of plutonium.

    a   How much plutonium is there in a tonne of uranium ore?

    b   In a year, 2 million tonnes of ore are mined.

    How much plutonium will this give?

(PS) **9**   Mr Ball was putting a path down his garden. He bought 30 paving slabs, each 0.75 m long. He wanted the path to be 21 m long.

    Has Mr Ball bought enough paving stones, or too many?

    How many extra or how many too few has be bought?

10 Max is in the DIY shop looking for some long nails. He sees that he can buy them at £0.03 each or for £1.75 for a box of 50. He wants to buy 80 nails.

What is the cheapest way for him to buy them?

Challenge

A Work out each multiplication.

   a 0.1×0.1 **b** $0.1 \times 0.1 \times 0.1$ **c** $0.1 \times 0.1 \times 0.1 \times 0.1$

B Using your answers to **A**, write down the answer to each of these multiplications by working it out in your head.

   a $0.1 \times 0.1 \times 0.1 \times 0.1 \times 0.1$

   b $0.1 \times 0.1 \times 0.1 \times 0.1 \times 0.1 \times 0.1$

   c $0.1 \times 0.1 \times 0.1 \times 0.1 \times 0.1 \times 0.1 \times 0.1$

   d $0.1 \times 0.1 \times 0.1 \times 0.1 \times 0.1 \times 0.1 \times 0.1 \times 0.1 \times 0.1 \times 0.1$

C Now write down the answer to each of these, by working them out in your head.

   a 0.2×0.2 **b** 0.3×0.3 **c** 0.4×0.4 **d** 0.5×0.5 **e** 0.8×0.8

   f $0.2 \times 0.2 \times 0.2$ **g** $0.3 \times 0.3 \times 0.3$ **h** $0.4 \times 0.4 \times 0.4$ **i** $0.5 \times 0.5 \times 0.5$

4.6 Dividing decimals

Learning objective

• To divide with decimals

When you divide with decimals:

• First change the number you are dividing by into a whole number, by multiplying it by 10, 100, 1000 as needed. Make sure that it is the smallest whole number possible. For example, change 0.2 to 2 by just multiplying by 10.

• Then multiply the number you are dividing into by the same amount.

Now you can complete the division as usual.

Example 13

Work these out.

a $0.08 \div 0.2$ **b** $20 \div 0.05$

   a You are dividing by 0.2 so multiply by 10 to make this into a whole number.

      Then multiply 0.08 by 10 as well, before you complete the division.

        $0.08 \div 0.2 = 0.8 \div 2 = 0.4$

   b Multiply 0.05 by 100 to make it into 5.

      Then multiply 20 by 100 as well, and divide.

        $20 \div 0.05 = 2000 \div 5 = 400$

If the dividing number is a whole number ending in zeros, you can make it easier to use by dividing by a multiple of 10. Then divide the other number by the same multiple of 10.

Example 14

Work these out.

a 4.8 ÷ 80   **b** 24 ÷ 3000

   a 4.8 ÷ 80

     Divide 80 by 10 to give 8.

     Then divide 4.8 by 10 as well, to give 0.48.

       4.8 ÷ 80 = 0.48 ÷ 8 = 0.06

     You can set the last step out as a short division problem if you want to, but be sure to keep the decimal points lined up.

$$\begin{array}{r} 0.06 \\ \hline 8)\overline{0.48} \end{array}$$

     So 4.8 ÷ 8 = 0.06.

   b 24 ÷ 3000 = 2.4 ÷ 300 = 0.24 ÷ 30 = 0.024 ÷ 3

$$\begin{array}{r} 0.008 \\ \hline 3)\overline{0.024} \end{array}$$

     So 24 ÷ 3000 = 0.008.

Exercise 4F

  1 Work these out without using a calculator.

    a 0.04 ÷ 0.02   **b** 0.8 ÷ 0.5   **c** 0.06 ÷ 0.1   **d** 0.9 ÷ 0.03

    e 0.2 ÷ 0.01   **f** 0.06 ÷ 0.02   **g** 0.09 ÷ 0.3   **h** 0.12 ÷ 0.3

    i 0.16 ÷ 0.2   **j** 0.8 ÷ 0.02   **k** 0.8 ÷ 0.1   **l** 0.24 ÷ 0.08

    m 0.2 ÷ 0.2   **n** 0.08 ÷ 0.8   **o** 0.9 ÷ 0.09   **p** 0.4 ÷ 0.001

  2 Work these out without using a calculator.

    a 200 ÷ 0.4   **b** 300 ÷ 0.2   **c** 40 ÷ 0.08   **d** 200 ÷ 0.02

    e 90 ÷ 0.3   **f** 40 ÷ 0.04   **g** 50 ÷ 0.1   **h** 400 ÷ 0.2

    i 300 ÷ 0.5   **j** 400 ÷ 0.05   **k** 400 ÷ 0.1   **l** 200 ÷ 0.01

    m 30 ÷ 0.5   **n** 50 ÷ 0.5   **o** 60 ÷ 0.5   **p** 400 ÷ 0.5

  3 Work these out without using a calculator.

    a 3.2 ÷ 20   **b** 2.4 ÷ 400   **c** 1.2 ÷ 400   **d** 3.6 ÷ 90

    e 24 ÷ 800   **f** 2.4 ÷ 2000   **g** 1.4 ÷ 70   **h** 1.6 ÷ 40

    i 32 ÷ 2000   **j** 0.18 ÷ 300   **k** 0.24 ÷ 200   **l** 0.032 ÷ 4000

4 Screws cost £0.03 each.

    How many can I buy with £6000?

4 Decimal numbers

5 A grain of salt weighs 0.002 g.

How many grains are there in a kilogram of salt?

6 A gallon of seawater contains about 0.000 002 kg of gold.

How many gallons will produce 3 kg of gold?

 7 A grain of sand weighs approximately 0.045 g.

Billy has a bucket that holds 2.5 kg of sand. Approximately how many grains are there in a full bucket of sand?

 8 At a market, Owen sees some brass hinges at a bargain price of £1.08 each. He has £90 in cash with him. What is the biggest number of hinges he can buy?

9 In this number wall, two numbers are multiplied to give the number above.

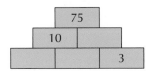

Work out the missing numbers to make the wall correct.

10 What is the largest possible integer you can divide into 333, that:

a leaves no remainder

b leaves a remainder

c gives a decimal answer between 10 and 11?

Challenge

A Given that $46 \times 34 = 1564$, write down the answer to:

   a 4.6×34 **b** 4.6×3.4 **c** $1564 \div 3.4$ **d** $15.64 \div 0.034$.

B Given that $57 \times 32 = 1824$, write down the answer to:

   a 5.7×0.032 **b** 0.57×32000 **c** 5700×0.32 **d** 0.0057×32.

C Given that $2.8 \times 0.55 = 1.54$, write down the answer to:

   a 28×55 **b** $154 \div 55$ **c** $15.4 \div 0.028$ **d** 0.028×5500.

Ready to progress?

I can add and subtract decimal numbers.
I can multiply and divide decimal numbers by 100, 1000, 10 000 and 100 000.
I can estimate answers and check if an answer is about right to one decimal place.
I can round off to one decimal place.

I can multiply and divide a decimal by a decimal.

Review questions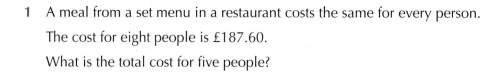

1 A meal from a set menu in a restaurant costs the same for every person.

   The cost for eight people is £187.60.

   What is the total cost for five people?

2 This table shows how much it costs to go to a local theatre.

	Matinee	Evening
Adult	£9.60	£12.90
Child (aged 14 or under)	£6.50	£11.50
Senior Citizen (aged 60 or over)	£7.35	£8.80

   Mrs Charlton (aged 35), her daughter (aged 8), her son (aged 6) and her mother (aged 64) want to go to the theatre.

   They are not sure whether to go the matinee or evening performance.

   How much will they save if they go to the matinee?

   Show your working.

3 Place these masses into order of size, writing the smallest first.

       2.05 kg   745 g   1.95 kg   3708 g   2 kg

4 Write down the next three numbers in each sequence.

   a   1.7, 4.4, 7.1, 9.8, …, …, …
   b   9 , 5.7 , 2.4, …, …, …
   c   1.2, 1.3, 2.5, 3.8, 6.3, …, …, …

5 Put these decimals in ascending order.

       60.8, 62.95, 6.37, 0.689, 6.4, 60.02, 6.13

6 Look at the shapes below. All lengths are in centimetres.

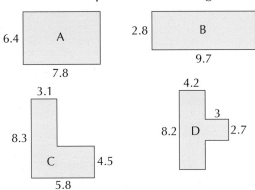

a Which shape has the longest perimeter?
b Which shape has the smallest area?

PS 7 A farmer sells geese for $10.55 each and pheasant for $16.40 each.

One market day he sells twice as many geese as pheasant and takes $9000.

How many of each did he sell?

PS 8 Gary is following a recipe for a chocolate cake and he needs 70 cm³ of milk chocolate. He has a thick bar of Belgian chocolate, with measurements as below.

a Gary estimates that he does not have enough chocolate in this bar. Show how he does this.
b Gary had another bar of chocolate that has the same width and depth but is 10 cm long.

He cuts it down to a length, in whole centimetres, that will give him just over 70 cm³ of chocolate. What is the length that he cuts?

9 A box of labels costs £17.40.

There are 75 sheets of labels in the box.

There are 12 labels on each sheet.

What is the cost of one label (to the nearest penny)?

PS 10 Trevor buys a 750 g box of salt. He says to his daughter: 'There are over a million grains of salt in this box – each grain weighs 0.065 g.'

His daughter, Vicky, told her dad he was wrong. Show that Vicky is correct.

Financial skills
Porridge is so good for you!

According to a popular survey, almost half the population in the UK start the day with porridge.

It is made with oats and is the perfect healthy breakfast food.

Some facts about the sales of porridge in the UK

☀ In the UK, 49 out of every 100 people start their day with porridge.

☀ Sales of porridge doubled in the last five years to £237 million.

☀ 38 out of every 100 young people eat porridge in the morning.

☀ About 138 000 hectares of land in the UK are used for producing oats.

☀ The UK produces 100 million tonnes of oats per year.

☀ Instant hot porridge pots cost 99p, you just need to add hot water.

☀ Ready porridge pots cost £1.49, you just put them in the microwave.

☀ A 1 kg box of oats costs £1.98.

☀ High-street coffee shops sell hot porridge for £2.32 per pot.

1 The UK population is about 63 million. How many of these people have porridge for breakfast?

2 What were the sales of porridge five years ago?

3 Rowlinson School has 1235 pupils. Approximately how many of them would you expect to have porridge for breakfast?

4 How many tonnes of oats are produced for every hectare in the UK?

5 Using 20 g of oats makes a good portion of porridge for one person. If a family of four had porridge for breakfast every day, how long would the 1 kg box of oats last them?

6 How much per day could the family of four save, if they use porridge made from the box rather than:

   a instant hot porridge pots where you just add hot water

   b ready porridge pots that you just put in the microwave?

7 High-street coffee shops take about £29 000 per day just from the sales of hot porridge. Approximately how many pots do they sell each day?

8 How much more would you pay for hot porridge from a high-street coffee shop than if you make it for yourself from the box at home?

5

Working with numbers

This chapter is going to show you:

- what square roots are
- how to use a calculator to work out square roots
- how to round to more than one decimal place
- how to round to one significant figure (1 sf)
- the order of operations
- how to carry out multiplication by written methods
- how to carry out division by written methods
- how to calculate with measurements.

You should already know:

- how to square a number
- multiplication tables up to 12×12
- how to round numbers to the nearest whole number, 10, 100 or 1000
- how to round to one decimal place (1 dp)
- the place value of digits in a number such as 23.5084
- how to carry out short multiplication and short division
- how to use a calculator to do simple calculations
- how to convert units of measurement.

About this chapter

Your school is organising its summer fête, but where should it locate the star attraction: a giant chessboard with an area of 9 m²? The organisers need to work out how long its sides are in order to see where it will fit. As the board is a square, its area is the length of one of its sides squared. This means that the length is the square root of the area. So if the area is 9 m² the sides will be 3 m long. Most square roots are not whole numbers, however. If the chessboard had an area of 10 m², the organisers would need either to approximate the square root or to use a calculator to find it, and then round it up, to be sure of having enough space to fit the board.

5.1 Square numbers and square roots

Learning objective

* To recognise and use square numbers up to 225 (15 × 15) and the corresponding square roots

Key words

power	square number
square root	squaring

When you multiply any number by itself, you are **squaring** the number. You can write it as a number to the **power** of 2, for example, $4 \times 4 = 4^2 = 16$.

When you multiply an integer (whole number) by itself, the result is called a **square number**.

You need to learn all of the first 15 square numbers.

1 × 1	2 × 2	3 × 3	4 × 4	5 × 5	6 × 6	7 × 7	8 × 8	9 × 9	10 × 10
1^2	2^2	3^2	4^2	5^2	6^2	7^2	8^2	9^2	10^2
1	4	9	16	25	36	49	64	81	100

11 × 11	12 × 12	13 × 13	14 × 14	15 × 15
11^2	12^2	13^2	14^2	15^2
121	144	169	196	225

The **square root** of a number is that number which, when squared, gives the starting number. For example:

* 13 is the square root of 169

* 14 is the square root of 196.

Taking a square root is the opposite of squaring a number.

A square root is represented by the symbol $\sqrt{}$. For example:

$\sqrt{1} = 1 \quad \sqrt{4} = 2 \quad \sqrt{9} = 3 \quad \sqrt{16} = 4 \quad \sqrt{25} = 5$

The square root of a number can be positive or negative. For example, -13 is also the square root of 169 because $-13 \times -13 = 169$.

You can write this as $\sqrt{169} = \pm 13$, which means $\sqrt{169} = +13$ or $\sqrt{169} = -13$.

Only the square root of a square number will give a whole number (integer) as the answer.

Example 1

Between which two consecutive positive whole numbers does $\sqrt{129}$ lie?

$11 \times 11 = 121$ and $12 \times 12 = 144$, so $\sqrt{129}$ is between 11 and 12.

Example 2

What possible answers could $\sqrt{9} + \sqrt{4}$ have?

Since $\sqrt{9} = \pm 3$ and $\sqrt{4} = \pm 2$, there are four possible answers.

* $3 + 2 = 5$

* $3 + -2 = 1$

* $-3 + 2 = -1$

* $-3 + -2 = -5$

Exercise 5A

Do not use a calculator unless the question tells you to do so.

(PS) 1 The two square numbers 64 and 225 add up to another square number, 289.
Find two different square numbers that add up to a square number.

(PS) 2 Find two square numbers that add up to 10^2.

3 Write down the value of each square root.

    a $\sqrt{225}$ **b** $\sqrt{100}$ **c** $\sqrt{169}$ **d** $\sqrt{121}$ **e** $\sqrt{196}$

4 Use a calculator to work out the value of each square root.

    a $\sqrt{256}$ **b** $\sqrt{2304}$ **c** $\sqrt{961}$ **d** $\sqrt{250000}$ **e** $\sqrt{576}$

    f $\sqrt{361}$ **g** $\sqrt{6889}$ **h** $\sqrt{4489}$ **i** $\sqrt{444889}$ **j** $\sqrt{44448889}$

(PS) 5 Each of these square roots lies between two positive consecutive whole numbers.
Work out what they are, in each case.

    a $\sqrt{40}$ **b** $\sqrt{95}$ **c** $\sqrt{60}$ **d** $\sqrt{190}$ **e** $\sqrt{220}$

    f $\sqrt{89}$ **g** $\sqrt{800}$ **h** $\sqrt{9000}$ **i** $\sqrt{5000}$ **j** $\sqrt{2015}$

6 Estimate the value of each square root. All the answers are whole numbers so use the units digit to help you. Then use your calculator to see how many of your estimates are correct.

    a $\sqrt{784}$ **b** $\sqrt{1225}$ **c** $\sqrt{2116}$ **d** $\sqrt{9801}$ **e** $\sqrt{2916}$

    f $\sqrt{961}$ **g** $\sqrt{1369}$ **h** $\sqrt{3025}$ **i** $\sqrt{2304}$ **j** $\sqrt{2401}$

(MR) 7 **a** Work out these square roots.

       i $\sqrt{121}$ **ii** $\sqrt{12100}$ **iii** $\sqrt{1210000}$

    b Write down what you notice about your answers to part **a**.

    c Use your answers to parts **a** and **b** to write down the value of $\sqrt{121000000}$.

8 What possible answers could $\sqrt{25} + \sqrt{49}$ have?

9 What possible answers could $\sqrt{169} - \sqrt{16} - \sqrt{9}$ have?

Investigation: Square numbers

A a Choose any three square numbers.

  b Multiply them together.

  c Work out the square root of the product.

  d Can you find a connection between this square root and the three starting numbers?

B Repeat part **A** for some more sets of three square numbers.

C Is the connection between the square root of the product and the three starting numbers the same, no matter what square numbers you choose?

D Is the connection the same if you use more than three square numbers?

5.2 Rounding

Learning objectives

- To round numbers to more than one decimal place (dp)
- To round numbers to one or two significant figures (sf)

Decimal places

You can already round decimal numbers to one **decimal place** (1 dp).

You can extend this, to round to two, three or more decimal places.

You do this in the same way as you round to one decimal place, by looking at the digit to the right of the one in the decimal place being rounded.

- If you are rounding to one decimal place, look at the digit in the second decimal place. If its value is 4 or less, **round down**. If its value is 5 or more **round up**.

- If you are rounding to two decimal places, look at the digit in the third decimal place. If it is 4 or less, round down. If it is 5 or more round up.

You can use this method to round to any number of decimal places.

You can use a wavy equals sign ($\approx$) to mean 'approximately equal to'. For example, if you round 34.57 to one decimal place, you could write $34.57 \approx 34.6$ which means '34.57 is approximately equal to 34.6'.

Example 3

Round each number to:

i one decimal place   **ii** two decimal places.

  a 8.715   **b** 5.451   **c** 28.967   **d** 32.991

  i Look at the hundredths digit (the second decimal place) and round it down or up.

  a $8.715 \approx 8.7$ **b** $5.451 \approx 5.5$ **c** $28.867 \approx 28.9$ **d** $32.991 \approx 33.0$
    ↑ round down   ↑ round up   ↑ round up   ↑ round up

  ii Look at the thousandths digit (the third decimal place) and round it down or up.

  a $8.715 \approx 8.72$ **b** $5.451 \approx 5.45$ **c** $28.867 \approx 28.87$ **d** $32.991 \approx 32.99$
    ↑ round up   ↑ round down   ↑ round up   ↑ round down

Significant figures

Sometimes, you will need to round numbers to a given number of **significant figures**.

- The digit with the highest place value in a number is the most significant figure.
- The digit with the next highest place value is the next most significant, and so on.

If you round a number to one significant figure (1 sf), it means you round it to its highest place value.

Look at these examples.

- In the number 382 649, the figure 3 has the highest place value, so it is the most significant. The figure 8 is the next most significant. If you rounded the number to one significant figure it would be 400 000.

- In the number 0.000 154 278 the figure 1 is the most significant, and the figure 5 is the next most significant. If you rounded the number to one significant figure it would be 0.0002.

Note: You do not include trailing zeros at the end of decimals when you round them to one significant figure, because 0.0002 is rounded to 1 sf. If you write it as 0.000 200 000, the five trailing zeros show that it has been written to six significant figures.

Example 4

Round each number to one significant figure.

a 9248   **b** 563   **c** 0.038 961   **d** 3.499

   a 9248 ≈ 9000   **b** 563 ≈ 600   **c** 0.038 961 ≈ 0.04   **d** 3.499 ≈ 3

      ↑ round down   ↑ round up   ↑ round up   ↑ round down

Note: In part **d**, 3.499 rounded to one significant figure is written as 3 and not 3.0 because 3.0 is taken as a number written to two significant figures.

Exercise 5B

1 Round each number to one decimal place (1 dp).

   a 17.51   **b** 11.67   **c** 41.13   **d** 74.25

   e 62.18   **f** 244.99   **g** 182.14   **h** 615.25

   i 181.05   **j** 61.41   **k** 108.09   **l** 9.99

2 Work out the answer to each calculation. Give your answers to one decimal place (1 dp).

   a 1.7×2.6 **b** 3.2×4.8 **c** 2.9×8.7 **d** $15.4 \div 9.5$

3 Work out the value of each square root. Give your answers to one decimal place (1 dp).

   a $\sqrt{40}$ **b** $\sqrt{95}$ **c** $\sqrt{60}$ **d** $\sqrt{190}$ **e** $\sqrt{220}$

   f $\sqrt{89}$ **g** $\sqrt{800}$ **h** $\sqrt{9000}$ **i** $\sqrt{5000}$ **j** $\sqrt{2015}$

4 These are the highest football league home crowd figures for a selection of teams in the English leagues.

Copy and complete the table.

Match (home fixture)	Attendance	Attendance (to 1 sf)
Arsenal	60 079	
Bristol City	26 575	
Cardiff City	37 933	
Carlisle United	14 530	
Exeter City	10 339	
Norwich City	26 672	
Manchester United	75 826	
Sheffield Wednesday	42 634	

5 Round each number to two decimal places (2 dp).

    a 18.623   **b** 1.7854   **c** 4.249   **d** 83.1455

    e 3.2943   **f** 2.8951   **g** 7.2425   **h** 5.795

    i 9.0459   **j** 7.5262   **k** 8.0963   **l** 9.999

6 Work out the answer to each calculation. Give your answers to two decimal places (2 dp).

    a $18.5 \div 2.3$ **b** $20.7 \div 1.9$ **c** $185 \div 11.2$ **d** $14.8 \div 8.65$

7 You can change a fraction into a decimal by dividing the numerator by the denominator, for example, $\frac{3}{8} = 3 \div 8 = 0.125$.

Change each fraction into a decimal and round the answer to one significant figure (1 sf).

    a $\frac{3}{7}$ **b** $\frac{5}{13}$ **c** $\frac{3}{25}$ **d** $\frac{19}{30}$

PS 8 The mass of the planets in our Solar system can be shown as a proportion of the mass of the Earth (MRE), as shown in the table.

Copy the table and write each number, correct to one significant figure, in the right-hand column.

	Mass relative to Earth (MRE)	MRE correct to 1 sf
Mercury	0.0553	
Venus	0.815	
Earth	1	
Mars	0.107	
Jupiter	318	
Saturn	95.2	
Uranus	14.5	
Neptune	17.2	

Problem solving: Rounding decimals

Joy and Chris are thinking of whole numbers.

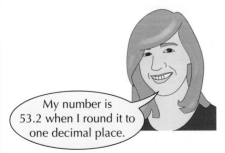

My number is 53.2 when I round it to one decimal place.

My number is 53.15 when I round it to two decimal places.

Chris's number is bigger than Joy's number.

Give an example to show that this could be true.

5.3 Order of operations

Learning objective

- To use the conventions of BIDMAS to carry out calculations

Key words

| BIDMAS | operation |
| order of operations | |

Most of the time, the order in which you carry out instructions is important.

These are the things that Kevin does when he is getting himself ready to go out in the morning.

Can you put them into the correct order?

Kevin can do some of these things in any order he chooses, but he must do some of them in the right order or he could look very odd!

In mathematics, the order in which calculations or **operations** are carried out is also important.

You can remember the **order of operations** by recalling **BIDMAS**.

B	Brackets
I	Indices or powers
D	Division
M	Multiplication
A	Addition
S	Subtraction

This means that you work out any brackets first, followed by any numbers with powers, then carry out division and multiplication, and finally addition and subtraction.

For example, in the calculation $9 + 7 \times 6$, you do the multiplication first.

$9 + 7 \times 6 = 9 + 42 = 51$

Example 5

Copy these calculations and circle the operation that you do first in each one.

Then work them out.

a $12 + 8 \div 2$ **b** $54 - 3 \times 7$ **c** $84 \div 12 - 9$ **d** $36 \div (29 - 11)$

  a Division is carried out before addition, so $12 + (8 \div 2) = 12 + 4 = 16$.

  b Multiplication is carried out before subtraction, so $54 - (3 \times 7) = 54 - 21 = 33$.

  c Division is carried out before subtraction, so $(84 \div 12) - 9 = 7 - 9 = -2$.

  d Brackets are carried out first, so $36 \div (29 - 11) = 36 \div 18 = 2$.

Example 6

Work these out, showing each step of the calculation.

a $11 + 13^2 \times 2 - 17$ **b** $(9 + 3)^2 + (15 - 2)^2$

a The order of working is power, multiplication, addition, subtraction (the last two can be interchanged). This gives:

power	$11 + 13^2 \times 2 - 17 = 11 + 169 \times 2 - 17$
multiplication	$= 11 + 338 - 17$
addition or subtraction	$= 349 - 17$
	$= 332$

b The order will be brackets (both of them), power, multiplication. This gives:

$$(9 + 3)^2 + (15 - 2)^2 = 12^2 + 13^2$$
$$= 144 + 169$$
$$= 313$$

If you have two equal operations, calculate in order, from left to right.

Example 7

Complete these calculations. **a** $7 - 8 - 3$ **b** $48 \div 12 \div 4$

a Working from left to right gives: $(7 - 8) - 3 = -1 - 3 = -4$

b Working from left to right gives: $(48 \div 12) \div 4 = 4 \div 4 = 1$

You can see that if you did these operations from right to left you would have different results.

 Hint You work through the steps in a calculation in the same order as you read them, left to right.

Example 8

Put brackets into each calculation, to make it true.

a $15 + 12 \times 4 = 108$ **b** $9 + 5^2 - 4 = 192$ **c** $234 \div 11 - 2 = 26$

Decide which operation is done first.

a $(15 + 12) \times 4 = 108$

b $(9 + 5)^2 - 4 = 192$

c $234 \div (11 - 2) = 26$

Exercise 5C

 1 Write down the operation that you do first in each of these calculations.
Then work each one out.

a $12 + 9 \times 8$ **b** $12 - 12 \div 3$ **c** $15 \times 15 + 20$ **d** $32 \div 8 - 6$

e $(12 + 13) \times 6$ **f** $(150 - 33) \div 3$ **g** $14 \times (11 + 3)$ **h** $72 \div (21 - 12)$

2 Work these out. Show all the steps in each calculation.

a $10 \times 8 + 12$ b $12 \times (3 + 4)$ c $12 + 13 \times 4$ d $(12 + 13) \times 4$

e $8 \times 8 - 8$ f $15 + 13^2 + 16$ g $15 \times (13^2 - 19)$ h $13^2 - (25 - 12)$

i $(12 + 13) \times (4 + 5)$ j $(12^2 + 6) \times (4 + 5)$ k $8 \div 8 + 8 \div 8$ l $88 \div 8 + 8$

m $(6 + 9)^2$ n $6^2 + 9^2$ o $11^2 + 9 \times 6$ p $(3 + 8) \times (9 - 12)$

 3 Add brackets to make each calculation true.

a $12 \times 5 + 7 = 144$ b $64 \div 8 \div 4 = 32$ c $7 + 5 \times 1 + 6 = 84$

d $9 + 6^2 \times 2 = 81$ e $3 + 12^2 = 225$ f $8 - 6 - 3 + 7 = 12$

g $3 \times 6 + 3 + 7 = 48$ h $5 + 4 \times 7 + 2 = 81$ i $19 - 34 - 12 = -3$

j $29 - 15 \times 2 = -1$ k $8 + 8 + 8 \div 2 = 12$ l $7 + 4^2 - 9 + 2 = 110$

 4 One of the calculations $2 \times 7^2 = 196$ and $2 \times 7^2 = 98$ is wrong.

a Which is it?

b Where could you add brackets to make it true?

5 Work these out.

a $(7 + 7) \div (7 + 7)$ b $(7 \times 7) \div (7 + 7)$ c $(7 + 7 + 7) \div 7$

d $7 \times (7 - 7) + 7$ e $(7 \times 7 + 7) \div 7$ f $(7 \times 7) - (7 \div 7)$

g $7 + 7 - 7 \div 7$ h $(7 + 7) \times (7 \div 7)$ i $(7 + 7) + 7 \div 7$

 6 Sophia is given four £10 notes and five £20 notes.

Write down the calculation you need to do to work out how much she is given altogether.

Work out the answer.

 7 Jan orders 12 garden plants at £2.99 each. Delivery costs £3.50

Write down the calculation you need to do to work out the total cost.

Work out the answer.

 8 Rashim wants to travel on four trams. He has €15.

The fares are €2.70, €4.20, €2.70 and €6.10

Does he have enough money?

Write down the calculation you need to do.

Work out the answer.

Problem solving: Sevens and eights

In Exercise 5C question 5, each calculation was made up of four 7s.

A Work out the value of: a $77 \div 7 - 7$ b $7 \times 7 - 7 \div 7$ c $7 \times 7 + 7 - 7$.

B Can you make up some other calculations, using four 7s, to give answers that you have not yet obtained in question 5 or in the three calculations above?

C See how many you can make, with values from 1 to 10.

D Repeat with four 8s. For example:

$(8 + 8) \div (8 + 8) = 1$ $8 \div 8 + 8 \div 8 = 2$

5.4 Multiplication problems without a calculator

Learning objective

- To use written methods to carry out multiplications involving decimals accurately

Key words

Chinese method

column method

grid or box method

long multiplication

Example 9

Helen was buying some books to use in the school library. She bought 36 books, each costing £4.30.

She had forgotten to take her calculator with her and needed to calculate the total accurately.

Show how she could have done this.

The problem is to work out $36 \times £4.25$.

Ignore the decimal point at first and work out 36×425.

There are a number of different methods Helen could use.

Grid or box method	Column method	Long multiplication	Chinese method

Grid or box method

×	30	6	
400	12 000	2400	14 400
20	600	120	720
5	150	30	180
			15 300

Column method

```
    425
  ×  36
     30   (6 × 5)
    120   (6 × 20)
   2400   (6 × 400)
    150   (30 × 5)
    600   (30 × 20)
  12000   (30 × 400)
  15300
```

Long multiplication

```
    425
  ×  36
   2550   (6 × 425)
  12750   (30 × 425)
  15300
```

Chinese method

$= 15 300$

Whichever method she uses, the answer to 36×425 is 15 300 and, as the original problem had two decimal places, she must put two decimal places back into the final answer, which makes it 153.00.

So, Helen paid £153.00 for the books.

Example 10

Brad bought 5.4 m of wood at a cost of £2.30 per metre.

How much did it cost him?

First, note there are three decimal places in the question.

Next, multiply 54 by 230 (any method you choose) to get 12 420.

Now put the three decimal places back into this result to make 12.420.

(A simple approximation gives $5 \times £2 = £10$.)

The final total will be £12.42.

1 Jonathan bought 23 model Dreamliner jets costing £9.88 each.

How much did this cost him?

2 A van travels an average of 8.7 miles to a litre of petrol. How many miles will the van travel on a full tank of 46 litres of petrol?

3 The school photocopier can print 92 sheets a minute.

If it runs without stopping for 4.8 minutes, how many sheets will it print?

4 A daily newspaper sells advertising by the square centimetre. On Monday, it sells 195 square centimetres at £1.85 per square centimetre. How much money does it get from this advertising?

(PS) 5 Tom was buying scout camping supplies at the supermarket. He wanted to buy 23 tins of meat. He saw some priced at £1.56 per tin and some others priced at £1.79 per tin. He didn't want to spend more than £40 on the meat. Could he buy 23 of either?

Show all your working.

(PS) 6 Kath has calculated that she needs 4.8 m² of carpet for her entrance hall.

When she was in the carpet showroom, she saw some carpet priced at £9.85 per square metre. She didn't want to spend over £40 on the carpet. Could she purchase enough of this carpet for less than £40?

(PS) 7 Len was walking down the street when he saw an advert.

He thought to himself: 'I need 3.75 kg of fertiliser. It looks as if I can get that for under £20.'

Is Len correct? Show your working.

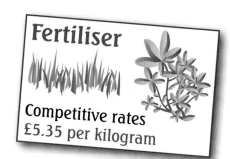

Fertiliser

Competitive rates
£5.35 per kilogram

8 The diagram shows the plan of a plot of land.

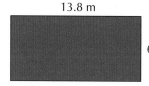

13.8 m

6.3 m

Calculate: **a** the area of the plot   **b** the perimeter of the plot.

(PS) 9 Work these out.

a 2.4 × 1.11   **b** 7.2 × 1.11   **c** 3.6 × 1.11

d What do you notice about the answers?

e Can you write down the answer to 1.7 × 1.11?

f Can you write down the answer to 9.5 × 1.11?

Investigation: Number patterns

A Copy and complete this number pattern as far as you can.

$(9 \times 9) + 7 =$

$(98 \times 9) + 6 =$

$(987 \times 9) + 5 =$

$(9876 \times 9) + 4 =$

B Write down a rule that describes the pattern you have found.

5.5 Division problems without a calculator

Learning objective

- To use written methods to carry out divisions involving decimals accurately

Key word

division

Sometimes you may need to solve a **division** problem when you have no calculator but only paper and pencil available, for you to use to work it out.

There are many different ways to work out a division calculation.

Example 11

Brian and five of his friends bought a lottery ticket together. They won £1098 between the six of them.

What will Brian's share be?

This can be calculated in a number of ways. Two methods are shown below.

Long division	Short division
183 6)1098 6 49 48 18 18 0	183 $6\overline{)10^{4}9^{1}8}$

The answer is £183.

Divisions do not always give exact answers, in whole numbers.

You will often need to add some decimal places and continue the division.

Example 12

Matt was at an auction without his calculator.

He bought eight identical bikes for a total of £970.

How much did he pay for each one?

Work out £970 ÷ 8.

Long division	Short division
$\begin{array}{r} 121 \cdot 25 \\ 8\overline{)970 \cdot 00} \\ \underline{8} \\ 17 \\ \underline{16} \\ 10 \\ \underline{8} \\ 20 \\ \underline{16} \\ 40 \end{array}$	$8\overline{)9^17^10 \cdot {}^20^40}$ $121 \cdot 25$

Notice that you need to write £970 as £970.00 to complete the division.

The answer is that the bikes cost £121.25 each.

Example 13

Peter has a roll of wire 12.9 m long.

He needs to cut as many pieces 0.8 m long as he can.

How many pieces can he cut?

Peter needs to calculate 12.9 ÷ 0.8.

Remember that when you divide with decimals you first make the dividing number into a whole number by multiplying it by 10, 100, 1000 as needed.

Make it the smallest number possible.

So multiply 0.8 by 10 in order to change it to a whole number.

$0.8 \times 10 = 8$

Then multiply the other number (12.9) in the division by the same amount.

$12.9 \times 10 = 129$

Now you can complete the division as usual.

$129 \div 8 = 16.125$

So Peter can cut 16 pieces.

Exercise 5E ✗

1. A family of five won £6316 in a lottery. They shared it equally among themselves. How much did each of them get?

2. Pete paid £43.47 to fill up his tank with 7 gallons of fuel. How much did he pay per gallon?

3. Gabriel has 94 litres of milk to pour equally into 8 containers. How many litres should he put into each one?

4. Lilly travels for 3 hours on a train, at a steady speed for 259 miles. How many miles did she cover each hour? Give your answer correct to one decimal place (1 dp).

(FS) 5. A group of 3250 people went on a journey from Paris to Rome. They travelled in 52-seater coaches.

   a How many coaches did they need?

   b Each coach cost €785. What was the total cost of the coaches?

   c How much was each person's share of the cost?

6. A contractor has 13 tonnes of waste to move by van. The van can carry 0.8 tonnes at a time.

   a How many trips must the van make to move all the waste?

   b A different contractor had a lorry that can take 1.1 tonnes as a time. How many trips would this contractor need to shift the 13 tonnes?

(PS) 7. The mathematics department has printed 400 sheets of questions on long division. They put them into sets of 30 sheets.

   How many more sheets do they need to print, so that there are no spare sheets after making up full sets?

(PS) 8. A local flour mill sells bags of 1.2 kg flour.

   In one day they can make 154 kg of flour.

   How many bags of flour can be filled from 154 kg of flour?

(PS) 9. To raise money, a running club is doing a relay race of 150 kilometres. Each runner except the last one will run 9 km. The last runner will run a shorter distance than the others.

   How far does the last one run?

(MR) 10. A pack of washing powder contains 2.75 kg of powder. The recommended amount to use per wash is 40 g of powder.

   How many washes can you get from the pack of washing powder?

Investigation: The number 37 037

37 037 is an interesting number.

Multiply 37 037 by various numbers and try to find what makes it such an interesting number. Explain your results clearly.

5.6 Calculations with measurements

Learning objective

- To convert between common metric units
- To use measurements in calculations
- To recognise and use appropriate metric units

You need to know and use these **metric conversions** for length and capacity.

Length	Capacity
1 kilometre (km) = 1000 metres (m)	
1 metre (m) = 100 centimetres (cm)	1 litre (l) = 100 centilitres (cl)
1 metre (m) = 1000 millimetres (mm)	1 litre (l) = 1000 millilitres (ml)
1 centimetre (cm) = 10 millimetres (mm)	1 centilitre (cl) = 10 millilitres (ml)

Can you see the connections?

- **Centi-** means hundredth, **cent-** means hundred, for example, a century is 100 years. There are 100 cents in a dollar.
- **Milli-** means thousandth, **mill-** means thousands, for example, a millennium is 1000 years.

You also need to know these metric conversions for mass.

Mass
1 kilogram (kg) = 1000 grams (g)
1 tonne (t) = 1000 kilograms (kg)

This table shows the relationship between the common metric units.

1000	100	10	1	$\frac{1}{10}$	$\frac{1}{100}$	$\frac{1}{1000}$
km		m			cm	mm
kg			g			mg
			l		cl	ml

Example 14

Convert:

a 8.5 centimetres to millimetres   **b** 275 grams to kilograms   **c** 5.2 litres to centilitres.

  a 1 cm = 10 mm   So multiply by 10.

     $8.5 \times 10 = 85$

     So 8.5 cm = 85 mm.

  b 1000 g = 1 kg   So divide by 1000.

     $275 \div 1000 = 0.275$

     So 275 g = 0.275 kg.

  c 1 litre = 100 cl   So multiply by 100.

     $5.2 \times 100 = 520$

     So 5.2 litres = 520 cl.

When adding or subtracting metric amounts that are given in different units, you need to **convert** then to the same unit first.

Example 15

Add together 84.13 m, 406 cm and 0.0581 km.

  First convert all the lengths to the same unit.

1000	100	10	1	$\frac{1}{10}$	$\frac{1}{100}$	$\frac{1}{1000}$
km			m		cm	mm
		8	4	1	3	
			4	0	6	
0	0	5	8	1		

  The answer is 0.146 29 km or 146.29 m or 14 629 cm.

  146.29 m is the sensible answer.

Example 16

Choose a sensible unit to measure each of these.

a The width of a football field   **b** The length of a pencil

c The mass of a car   **d** A spoonful of medicine

  Choose a sensible unit. Sometimes there is more than one answer.

  a Metre   **b** Centimetre   **c** Kilogram   **d** Millilitre

Exercise 5F

1 Convert each length to centimetres.
  a 80 mm   **b** 3 m   **c** 797 mm   **d** 0.008 km   **e** 31.75 m

2 Convert each length to kilometres.
  a 236 m   **b** 8045 m   **c** 6002 cm   **d** 21 118 mm   **e** 76 m

3 Convert each length to millimetres.
  a 84 cm   **b** 3.2 m   **c** 0.41 km   **d** 25.6 cm   **e** 0.82 cm

4 Convert each mass to kilograms.
  a 5759 g   **b** 817 g   **c** 19 430 g   **d** 82 g   **e** 760 g

5 Convert each mass to grams.
  a 6.5 kg   **b** 5.43 kg   **c** 0.87 kg   **d** 0.003 kg   **e** 5.623 kg

6 Convert each capacity to litres.
  a 128 cl   **b** 2099 ml   **c** 2942 cl   **d** 36 cl   **e** 85 285 ml

7 Convert each time to hours and minutes.
  a 72 minutes   **b** 325 minutes   **c** 870 minutes   **d** 400 minutes   **e** 93 minutes

8 Add together the measurements in each group and give the answer in an appropriate unit.
  a 4.08 m, 86 cm, 0.0071 km   **b** 0.305 kg, 64 g, 0.006 kg
  c 8.5 litres, 42 cl, 936 ml   **d** 0.0007 km, 470 mm, 3.54 cm

(MR) 9 Pierre buys 1 kg of sugar, 620 grams of bananas and 1.28 kg of apples.
  Is the total mass of the three items more than 3 kilograms?
  Show your working.

(PS) 10 Fill in each missing unit.
  a A two-storey house is about 8 . . . high.   **b** Jon's mass is about 53 . . . .
  c Rob lives about 3 . . . from school.   **d** Luke ran a marathon in 3.5 . . . .

11 Read the value on each scale.
  a Give your answer in grams.

b Give your answer in kilograms.

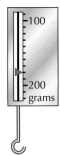

c Give your answer in kilograms.

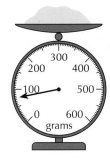

12 Read the value from each scale.

Give your answers correct to one significant figure (1 sf).

a
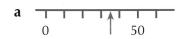

b

c

(PS) **13** Rachel has to take 5 ml of medicine four times a day. She must take this for 60 days.
How many litres of the medicine should she be given?

Investigation: Growing plants in rows

A A farmer grows plants in rows 1 metre apart. The plants are 1 m apart in each row.
How many plants could he grow on a plot of land 2500 m² in area?

B There is more than one possible answer to the above problem.
Try to show a few different possible answers.

Ready to progress?

  I know and can use square numbers up to 15×15 and the corresponding square roots.

  I can round up or down to one decimal place.
I can use the correct order of operations to carry out calculations.
I can use written methods to solve problems involving multiplications and divisions.
I can convert units of measurements and calculate with them.

  I can use written methods to carry out multiplications and divisions involving decimals.
I can round up or down to two decimal places.
I can round to one significant figure.

Review questions

1   Here is a sequence of numbers.

    $1, -3, 9, -27, \ldots$

    a   Write down the next three numbers in the sequence.
    b   Write down the rule you used to work out the next numbers.

2   What is the perimeter of this rectangle?

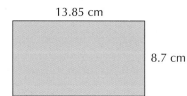

13.85 cm

8.7 cm

 3   Work these out.

    a $-12 + 7 \times 4$ b $(-4 - 9) \times (3 - 7)$
    c $(-6)^2 - (-4)^2 \div 8$

4   Which of these shapes has the greater area?

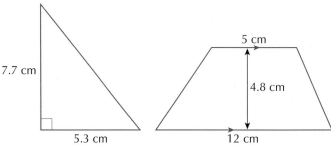

7.7 cm

5.3 cm

5 cm

4.8 cm

12 cm

    You must show all your working.

5   Explain the difference between rounding to one decimal place and rounding to one significant figure.

6 Kamil is in a superstore without his calculator. He is looking at hockey shirts and shorts. He needs to buy seventeen of each.

He sees the following offer.

Hockey shirts		Hockey shorts	
Basic	$7.35 each	Basic	$4.15 each
Stylish	$9.88 each	Stylish	$7.92 each

Kamil only has $300 to spend.

The shirts must all be of the same type, and the shorts must all be of the same type.

The shirts and the shorts can each be either basic or stylish.

What is the most expensive combination he can buy?

7 In a wildlife park, the young giraffe weighed three times more than the adult penguin.

Together they weigh 210 kg.

How much does the baby giraffe weigh?

8 Every second, 200 cm³ of water comes out of a tap into a cuboid water tank.

The base of the tank is 25 cm by 35 cm. The height is 10 cm. How long does it take to fill the tank? Give your answer correct to one decimal place.

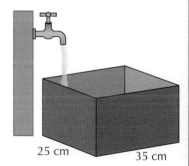

9 In this brick wall, the product of two numbers that are next to each other gives the number in the brick above. Copy the wall and and write down the numbers to complete it.

-1.3	4	-2.1

Problem solving

What is your carbon footprint?

Fascinating facts

- The average yearly carbon dioxide emissions in the UK are 9.5 tonnes per person.

- A family car emits about 160 g of carbon dioxide per kilometre.

- A small car emits about 100 g of carbon dioxide per kilometre.

- A large 4 by 4 emits about 300 g of carbon dioxide per kilometre.

- A lorry emits about 0.169 kg of carbon dioxide per mile, for every tonne of cargo it carries.

- An aeroplane emits about 0.806 kg of carbon dioxide per mile, for every tonne of cargo it carries.

- A boat emits about 0.041 kg of carbon dioxide per mile, for every tonne of cargo it carries.

Carbon dioxide emissions per person

1 A rule for working out the average yearly carbon dioxide emissions of people in Australia is to double the UK figure and add 1. Work out the average yearly carbon dioxide emissions for someone in Australia.

2 A rule for working out the average yearly carbon dioxide emissions of people in Brazil is to divide the UK figure by 3 and subtract 1. Work out the average yearly carbon dioxide emisions for someone in Brazil.

Carbon dioxide while we travel

3 Work out the carbon dioxide emission for small car on a journey of of 16 km.

4 Work out the carbon dioxide emission for a family car on a journey of 16 km.

5 A family of six all travel together in a large 4 by 4 on a journey of 16 km.

What is the carbon dioxide emission per person for the journey?

Bringing food to the UK

6 20 kg of strawberries are flown from Turkey to Southampton (1760 miles), then taken by road to Birmingham (133 miles).

   a Work out the carbon dioxide emission for the strawberries on this journey.

   b On average, one strawberry weighs 16 g. What is the carbon dioxide emission for each strawberry on this journey?

7 200 kg of tomatoes are brought by boat from Mexico to Southampton (5551 miles), then taken by road in a lorry to Bristol (78 miles).

   a What is the carbon dioxide emission for the tomatoes on this journey?

   b On average, one Mexican tomato weighs 55 g. What is the carbon dioxide emission for each tomato on this journey?

8 750 kg of peas are brought by road from Doncaster to London (172 miles).

   a What is the carbon dioxide emission for the peas on this journey?

   b The average mass of one pea is 7 g. What is the carbon dioxide emission for each pea on this journey?

6

Statistics

This chapter is going to show you:

- how to calculate the mode, the median, the mean and the range for a set of data
- how to use the assumed mean
- how to interpret statistical diagrams and charts
- how to use tally charts and frequency tables to collate data
- how to collect and organise discrete and continuous data
- how to create data collection forms and questionnaires
- how to draw simple conclusions from data.

You should already know:

- how to interpret data from tables, graphs and charts
- how to draw line graphs, frequency tables and bar charts
- how to create a tally chart
- how to draw bar charts and pictograms.

About this chapter

How many people are there in the world? Or even in our country? How do they live? What do they eat and drink? How big are their families?

We find out statistics like these by carrying out censuses and surveys. Censuses are huge surveys that find out information about every single man, woman and child in a country.

In the UK a census is carried out every 10 years. When the data is analysed and interpreted it helps the government decide what it needs to do. Charts and graphs give us tools for analysing and representing statistical data crucial for drawing the right conclusions from it.

6.1 Mode, median and range

Learning objective

- To understand and calculate the mode, median and range of data

Statistics is all about collecting and organising data, then using diagrams to represent and interpret it so that you can understand what it might mean.

You often need to find an **average** to help you to interpret data.

An average is a single or typical value that represents a whole set of values. This makes it easier to understand what the data shows.

This section explains how to find two types of average: the **mode** and the **median**:

- The mode is the value that occurs most often in a set of data. It is the only average that you can use for non-numerical data.

  Sometimes there may be no mode because either all the values are different, or no single value occurs more often than any other values.

- The median is the middle value for a set of values when they are put in numerical order.

The **range** of a set of values is the difference between the largest and smallest values.

This is equal to the largest value minus the smallest.

A small range means that the values in the set of data are similar in size, whereas a large range means that the values differ a lot and therefore are more spread out.

Example 1

These are the ages of 11 players in a football squad.

23, 19, 24, 26, 27, 27, 24, 23, 20, 23, 26

Find:

a the mode

b the median

c the range.

a First, put the ages in order.

19, 20, 23, 23, 23, 24, 24, 26, 26, 27, 27

The mode is the number that occurs most often.

So the mode is 23.

b The median is the number in the middle of the set.

This will be the sixth of 11 values.

So the median is 24.

c The range is the largest number minus the smallest number.

$27 - 19 = 8$

So the range is 8.

Example 2

These are the marks of 10 pupils in a mental arithmetic test.

19, 18, 16, 15, 13, 14, 20, 19, 18, 15

Find: **a** the mode   **b** the median   **c** the range.

First, put the marks in order.

13, 14, 15, 15, 16, 18, 18, 19, 19, 20

a There are three modes, as 15, 18 and 19 each appear twice.

b There are two numbers in the middle of the set: 16 and 18.

The median is the number that would be in the middle of these two numbers.

So the median is 17.

c The range is the largest number minus the smallest number.

20 − 13 = 7

So the range is 7.

Exercise 6A

1 Find the median and the range of the following sets of data.

    a   7, 6, 2, 3, 1, 9, 5, 4, 8

    b   36, 34, 45, 28, 37, 40, 24, 27, 33, 31, 41

    c   14, 12, 18, 6, 10, 20, 16, 8, 5

    d   99, 101, 107, 103, 109, 102, 105, 110, 100, 98, 99

    e   23, 37, 18, 23, 28, 19, 21, 25, 36

    f   3, 1, 2, 3, 1, 0, 4, 2, 4, 2, 2, 6, 5, 4, 5

    g   2.1, 3.4, 2.7, 1.8, 2.2, 2.6, 2.9, 1.7, 2.3

    h   2, 1, 3, 0, −2, 3, −1, 1, 0, −2, 1

2 Find the mode, median and range of each set of data.

    a   £2.50, £1.80, £3.65, £3.80, £4.20, £3.25, £1.80

    b   23 kg, 18 kg, 22 kg, 31 kg, 29 kg, 32 kg

    c   132 cm, 145 cm, 151 cm, 132 cm, 140 cm, 142 cm

    d   32°, 36°, 32°, 30°, 31°, 31°, 34°, 33°, 32°, 35°

3 A group of nine Year 7 pupils had their lunch in the school cafeteria.

These are the amounts that they spent.

    £2.30, £2.20, £2.00, £2.50, £2.20, £2.90, £3.60, £2.20, £2.80

    a   Find the mode for the data.

    b   Find the median for the data.

    c   Which is the better average to use? Explain your answer.

4 **a**   Write down a list of seven numbers with a median of 12 and a mode of 10.

    b   Write down a list of eight numbers with a median of 12 and a mode of 10.

    c   Write down a list of seven numbers with a median of 12, a mode of 11 and a range of 6.

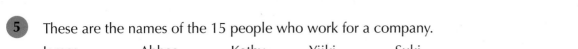

5 These are the names of the 15 people who work for a company.

James	Abbas	Kathy	Yiiki	Suki
Lucy	Brian	Kathy	Lucy	Tim
Bernard	Kathy	James	Ryan	Tim

One person leaves the company. A different person joins the company.

Now the name that is the mode is Tim.

a What is the name of the person who leaves?

b What is the name of the person who joins?

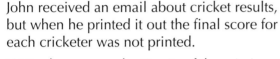 **6** **a** There are three children in the Bishop family. The range of their ages is exactly 6 years. What could the ages of the three children be? Give an example.

b There are two children in the Patel family. They are twins. What is the range of their ages?

7 John received an email about cricket results, but when he printed it out the final score for each cricketer was not printed.

Write down a good estimate of the missing number for each list. Give your reasons.

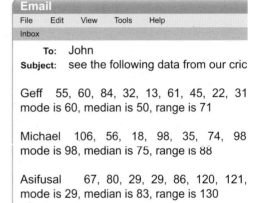

Email

File   Edit   View   Tools   Help

Inbox

To: John
Subject: see the following data from our cric

Geff   55, 60, 84, 32, 13, 61, 45, 22, 31
mode is 60, median is 50, range is 71

Michael   106, 56, 18, 98, 35, 74, 98
mode is 98, median is 75, range is 88

Asifusal   67, 80, 29, 29, 86, 120, 121,
mode is 29, median is 83, range is 130

8 At Greystones School sports day two scorers recorded the finishing times for different runners in the 800 metres.

They recorded them in different ways. These were the times for the five runners.

6 minutes 25 seconds, 6.25 minutes,
6.8 minutes, 7 minutes 22 seconds,
7.25 minutes

Find the median time taken for the race.

Activity: Mode, median and range

Find the mode, the median and the range of:

A pens and pencils you have

B the widths of the coins in your pocket

C the lengths of each of your fingers from the knuckle to the end of the fingernail. (Include the thumbs too.)

D the numbers of letters in the names of your friends.

6.2 The mean

Learning objective

- To understand and calculate the mean average of data

The **mean** is the most commonly used average. You may see it called the **mean average** or simply the **average** but, for clarity, it is better to call it the mean. The mean can be used only with numerical data.

The mean of a set of values is the sum of all the values divided by the number of values in the set.

$$\text{Mean} = \frac{\text{sum of all values}}{\text{number of values}}$$

The mean is a useful statistic because it takes all of the values into account, but it can be distorted by an **outlier.** This is a value, in the set of data, that is much larger or much smaller than the rest. When there is an outlier, the median is often used instead of the mean.

Example 3

Find the mean of 2, 7, 9, 10.

The mean is $\dfrac{2 + 7 + 9 + 10}{4} = \dfrac{28}{4} = 7$

For more complex data, you can use a calculator. When the answer is not exact, the mean is usually given to one decimal place (1 dp).

Example 4

The ages of seven people are 40, 37, 34, 42, 45, 39, 35. Calculate their mean age.

The mean age is $\dfrac{40 + 37 + 34 + 42 + 45 + 39 + 35}{7} = \dfrac{272}{7} = 38.9$ (1 dp)

If you don't have a calculator, you may be able to work out the mean more quickly by using an **assumed mean**, as the next example shows.

Example 5

Use the assumed mean method to find the mean of 146, 147, 137, 141, 139.

First, make an initial estimate of the mean. It needs to be a central value but does not have to be a value in the list.

Take the assumed mean for this list to be 140.

Then find the difference between each value in the list and the assumed mean.

These differences are 6, 7, −3, 1, −1.

The mean of the differences is: $\dfrac{6 + 7 + (-3) + 1 + (-1)}{5} = \dfrac{10}{5} = 2$

 Hint This answer may be negative for some examples.

The actual mean is the assumed mean + the mean of the differences.

So, the actual mean for the list of numbers is 140 + 2 = 142.

Exercise 6B

Use a calculator unless the question tells you not to.

1 Calculate the mean of each set of data.

   a 9, 8, 7, 11, 5   **b** 26, 35, 43, 40, 32, 28

   c 22, 23, 20, 37, 25, 28, 27   **d** 1.3, 0.5, 2.1, 0.7, 3.1, 1.4, 3.4, 1.1

2 Calculate the mean of each set of data, giving your answers correct to 1 dp.

   a 7, 8, 7, 5, 2, 4   **b** 13, 16, 18, 11, 19, 17, 14

   c 81, 75, 85, 98, 50, 70, 80, 83   **d** 9.4, 8.1, 10.6, 8.8, 11.9, 9.2

3 Do not use a calculator for this question.

Use the assumed mean method to find the mean of each set of data.

   a 27, 32, 39, 34, 26, 28

   b 97, 106, 89, 107, 98, 104, 95, 104

   c 237, 256, 242, 251, 238, 259, 245, 261, 255, 236

   d 30.6, 29.8, 31.2, 28.7, 32.8, 29.3, 31.8

4 These are the heights, in centimetres, of 10 children.

     129, 144, 139, 133, 132, 143, 150, 129, 134, 146

   a Calculate the mean height of the children.

   b What is the median height of the children?

   c What is the modal height of the children?

   d Which average do you think is the best one to use? Explain your answer.

5 These are the numbers of children in the families of Dan's class at school.

     1, 1, 1, 1, 1, 2, 2, 2, 2, 3, 3, 3, 3, 3, 3

   a What is the mode?

   b What is the median?

   c What is the mean?

6 These are the shoe sizes of all the boys in class 7JS.

     4, 4, 4, 4, 4, 5, 5, 5, 6, 7, 7

   a What is the mode?

   b What is the median?

   c What is the mean?

7 These are the weekly wages of 12 office staff in a small company.

     £150, £175, £135, £500, £130, £450, £180, £250, £150, £150, £155, £175

   a Calculate the mean weekly wage of the staff.

   b How many staff earn more than the mean wage?

   c Explain why so few staff earn more than the mean wage.

8 Tom goes on a five-day walking holiday.

The table shows how far he walked on the first four days.

Monday	Tuesday	Wednesday	Thursday
15 km	24 km	14 km	14 km

Tom says: 'My average for the first four days is more than 16 km.'

a Explain why Tom is correct.

Friday is his last day walking and he wants to increase his average to 18 km.

b How many kilometres must he walk on Friday?

Investigation: Mean and median

A Vital statistics

Calculate the mean and the median for your group's handspans and the circumferences of your heads.

B Average score

a Throw a dice 10 times. Record your results. What is the mean score?

b Repeat the experiment, but throw the dice 20 times. What is the mean score now?

c Repeat the experiment, but throw the dice 50 times. What is the mean score now?

d Write down anything you notice as you throw the dice more times.

6.3 Statistical diagrams

Learning objective

Key words	
bar chart	grouped data
line graph	pie chart

• To be able to read and interpret different statistical diagrams

When you have collected data from a survey, you can display it in various ways, to make it easier to understand and interpret.

The most common ways to display data are **bar charts**, **pie charts** and **line graphs**.

• Bar charts may be drawn in several different ways. You can show data that has single categories in a bar chart with gaps between the bars. For **grouped data**, for example, where data values fall into ranges such as 1–5, 6–7, you use a bar chart with no gaps between the bars.

• Pie charts show data that falls into several categories. A pie chart shows what fraction of the whole set of data each category represents.

• You would generally use line graphs to show trends and patterns in the data. They often show what happens over time.

When you compare charts of data from different sources it is important to be sure that they are based on similar numbers. This applies especially to pie charts, which show proportions but do not usually show the numbers in each category.

For example, two different classes did a survey of how the pupils travelled to school. They both made pie charts to show their results.

Jim's class

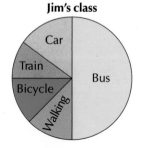

Noriko's class

In which class did the higher number of pupils use the bus? It seems obvious that it is Jim's class. But if you are told that there are 24 pupils in Jim's class and 36 in Noriko's you will see that:

• in Jim's class, half of the 24 pupils used the bus, which is a total of 12

• in Noriko's class, more than one third of the 36 pupils used the bus, which is a total of about 14.

Here, you can say that a bigger proportion of Jim's class used the bus, but you cannot say that more pupils in Jim's class used the bus than in Noriko's class.

Exercise 6C

1 The bar chart shows how the pupils in a class travel to school.

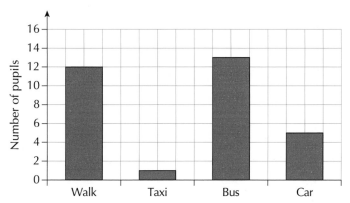

a How many pupils walk to school?

b What is the mode for the way the pupils travel to school?

c How many pupils are there in the class?

2 The dual bar chart shows the average monthly rainfall, in millimetres, for England and Wales over two separate decades, 1991–2000 and 2001–2010.

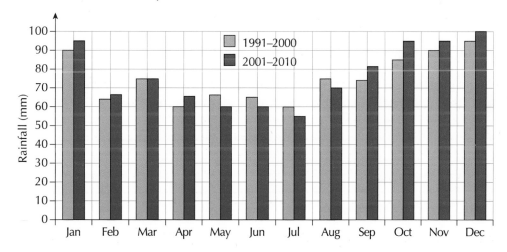

a Which month is always the wettest?

b Which months had increased rainfall in 2001–2010?

c What was the range of rainfall in each decade?

d Describe the changing weather pattern between the two decades.

3 The line graph shows the temperature, in Celsius degrees (°C), in Sheffield over an 18-hour period.

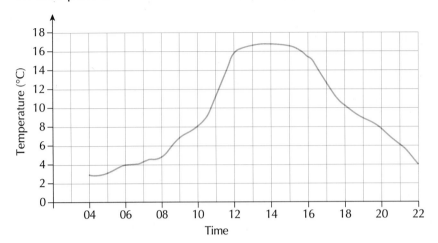

a What was the temperature at midday?

b What was the temperature at 8:00 am?

c Write down the range for the temperature over the 18-hour period.

d By listing hourly readings, calculate the mean temperature for the day.

e Explain why the line graph is a useful way of showing the data.

4 This compound bar chart shows the colours of the teachers' cars in the car park one day.

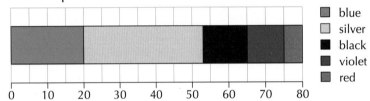

- blue
- silver
- black
- violet
- red

a Which is the modal colour of the teachers' cars?

b How many black cars were there in the car park?

c What percentage of the teachers had blue cars?

d Explain why the compound bar chart is a useful way to illustrate the data.

 5 The bar chart shows the monthly car sales of a firm in England during one year.

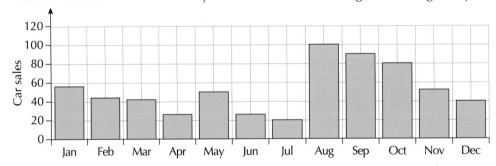

a What is the difference in sales between the first half and the second half of the year?

b In which month are the car sales the highest?

c What is the range of the monthly sales?

d What percentage of cars were sold in the month of August?

6   Nine hundred pupils in a school were questioned about which subjects they enjoyed the most. The pie chart displays their responses.

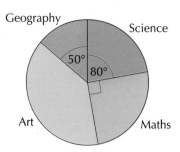

Geography

Science

50°

80°

Art

Maths

a   How many enjoyed Science most?

b   What fraction enjoyed Maths most?

c   What percentage enjoyed Geography the most?

Activity: Statistical diagrams in the press

A Look through newspapers and magazines and find as many statistical diagrams as you can. Make a display to show the variety of diagrams used in the press.

B What types of diagram are most common?

C How effective are the diagrams in showing the information?

D Are any of the diagrams misleading? If they are, explain why.

6.4 Collecting and using discrete data

Learning objective

• To create and use a tally chart

Key words

discrete data	frequency
modal	sample
tally chart	

What method of transport do pupils use to travel to school?

If you ask pupils this question, they will name different methods of transport, such as bus, car, bike, walking, train and they will give various reasons for using them!

A good way to collect this data is to fill in a **tally chart** as each pupil responds to the question. For example:

Type of transport	Tally	Frequency
Bus	⾏Ⅲ IIII	9
Car	⾏Ⅲ	5
Bike	II	2
Walking	⾏Ⅲ ⾏Ⅲ IIII	14
Other		
	Total:	30

Each 'stick' represents one pupil. When you get to the fifth, you draw it as a sloping stick across the other four. Using tallies allows you to collect the data and count it easily in fives.

The **frequency** is the sum of all the sticks in the tally.

You can see here that the most common type of transport is walking. This is the mode of the data and so you can say the **modal** form of transport is walking.

This sort of data is called **discrete data** because there is only a fixed number of possible answers.

Discrete data can be presented in bar charts, with gaps between the bars, or as pie charts. Sometimes collecting every single piece of data would be very time-consuming. How would you find the answer to this question?

Do certain newspapers use more long words than the other newspapers?

The most accurate way would be to count all of the words, in each newspaper, and compare the numbers of letters in the words. Because this would take far too long, instead you would take a **sample**. You would count, say, 100 words from each newspaper and find the length of each of these words.

Exercise 6D

1 The pupils in a class were asked: 'Where would you like to go for your form trip?'

This is how they voted.

a Draw a chart illustrating the places the pupils wanted to go to.

b Write down some suitable reasons why the pupils might have voted for each place.

c What is the modal place chosen?

Place	Tally
Museum	⾏Ⅲ ⾏Ⅲ
Stonehenge	⾏Ⅲ I
The zoo	⾏Ⅲ III
A circus	III
Theatre	IIII

2 The pupils in a class were asked: 'What is your favourite wild animal?'

This is how they voted.

    a Draw a chart illustrating their favourite wild animals.

    b Write suitable reasons why the pupils might have voted for each animal.

    c What is the modal animal chosen?

Animal	Tally
Snow leopard	卌 ||
Lion	卌 ||||
Giraffe	卌
Elephant	||||
Other	|||

3 The pupils in a class were asked: 'What is your favourite Olympic sport?'

This is how they voted.

    a Draw a chart illustrating their favourite sports.

    b Write suitable reasons why the pupils might have voted for each sport.

    c What is the modal sport chosen?

Sport	Tally
Running	卌 |
Cycling	卌
Swimming	卌 |||
Horse-riding	卌 ||
Other	||||

4 The pupils in a class were asked: 'What is your favourite school subject?'

This is how they voted.

    a Draw a chart illustrating their favourite subjects.

    b Write down suitable reasons why the pupils might have voted for each subject.

    c What is the modal subject chosen?

Subject	Tally
Spanish	卌
Biology	||||
Maths	卌 卌 |
PE	卌 |
Physics	|||
Other	|

5 Use your own class tally sheet to draw a chart illustrating the methods of transport used by pupils to get to school, and the reasons why.

Activity: How many letters

A Select one or two pages from a book.

B Create a data capture form (tally chart) like the one below.

Number of letters	Tally	Frequency
1		
2		
3		
4		
5		

C a Select at least two different paragraphs from different chapters.

   b Count the letters in each word and complete the tally.

    Note:

- numbers such as 3, 4, 5 count as one letter
- numbers such as 15, 58 count as two letters
- numbers such as 156, 897 count as three letters, and so on
- for hyphenated words, such as vice-versa, ignore the hyphen and count as two separate words.

D Fill in the frequency column.

E Create a bar chart for your results.

F What is the modal number of letters?

G You may find it interesting to compare the differences between the different books the class has worked on.

6.5 Collecting and using continuous data

Learning objective

- To understand continuous data and use grouped frequency

Key words

continuous data

class

grouped frequency

grouped frequency table

modal class

How far do you travel to school?

Pupils could give any answer to this question within a range, such as 2 miles, 1.3 miles, 1.25 miles, so data such as this is called **continuous data**.

Every item of data you collect may be different, so how will you make sense of it?

A good way to do this is to put it into groups.

For example, you could put replies of 2 miles, 1.3 miles, 1.25 miles into a group of 'more than 0 miles but equal to or less than 2 miles'. These groups are called **classes** and they are described by using the symbols < and ≤.

So 2 < distance ≤ 4 means the distance is more than 2 miles and less than or equal to 4 miles.

This class will include distances of 4 miles but not 2 miles.

Distance to travel to school (miles)	Tally	Frequency
0 < distance ≤ 2	ЦНТ ‖	7
2 < distance ≤ 4	‖‖	4
4 < distance ≤ 6	‖‖	3
6 < distance ≤ 8	ЦНТ ЦНТ ‖‖	13
	Total	27

How long does it take you to get to school in the morning?

When a class was asked this question, these were their replies.

6 minutes	3 minutes	5 minutes	20 minutes	15 minutes	11 minutes
13 minutes	28 minutes	30 minutes	5 minutes	2 minutes	6 minutes
8 minutes	18 minutes	23 minutes	22 minutes	17 minutes	13 minutes
4 minutes	2 minutes	30 minutes	17 minutes	19 minutes	25 minutes
8 minutes	3 minutes	9 minutes	12 minutes	15 minutes	8 minutes

There are too many different values in this list to make a sensible bar chart.

You need to group them, to produce a **grouped frequency table**, as shown below. The data has been put into different groups, called classes. Where possible, classes are kept the same size as each other.

Time (minutes)	0–5	6–10	11–15	16–20	21–25	26–30
Frequency	7	6	6	5	3	3

Notice that 6–10, 11–15, … mean 'over 5 minutes up to 10 minutes', 'over 10 minutes up to 15 minutes' and so on.

A frequency diagram drawn from this data looks like this.

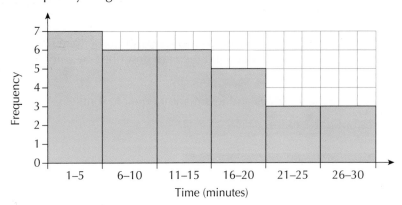

You cannot find a mode for grouped data. Instead, you use the **modal class**. This is the class with the highest frequency.

However, the bar chart shown above does not say what has happened to the in-between times, from 5 minutes to 6 minutes, from 10 to 11 minutes, and so on. The times could have been rounded to the nearest minute.

Alternatively, you could use a continuous scale for this data.

Time to travel to school (minutes)	Frequency
0 < time ≤ 5	7
5 < time ≤ 10	6
10 < time ≤ 15	6
15 < time ≤ 20	5
20 < time ≤ 25	3
25 < time ≤ 30	3

The diagram would now look like this.

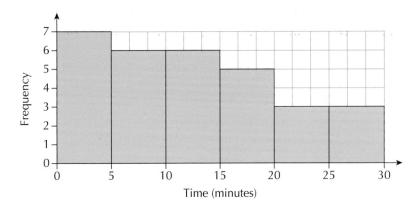

Exercise 6E

1 A year group did a survey on how many text messages each pupil had sent the day before. These are the results of this survey.

5	8	3	19	2	17	20	16	14	0	10	18
5	7	11	13	16	9	4	15	3	15	16	19
6	17	4	7	6	19	13	6	10	20	6	18
18	17	6	9	8	8	11	17	18	11	8	20
3	17	16	19	7	6	8	9	2			

a Create a grouped frequency table with a class size of 5, as below.

Number of texts	Tally	Frequency
0–4		
5–9		
10–14		
15–19		
20–24		
	Total:	

b Use the data above to complete the table.

c Draw a bar chart of the data.

d What is the modal class?

2 A teacher asked her year group: 'Count how many times this week you play electronic games.'

These were their responses.

5	8	10	5	25	19	9	10	30	29	4	2
1	7	22	26	15	22	8	6	26	11	9	2
1	27	27	11	15	19	16	18	6	10	13	8
29	24	7	11	14	30	8	9	1	8	24	29
27	13	5	26	14	13	8	29				

a Create a grouped frequency table with a class size of 5, as below.

Number of times	Tally	Frequency
1–5		
6–10		
11–15		
16–20		
21–25		
26–30		
	Total:	

b Use the data above to complete the table.

c Draw a bar chart of the data.

d What is the modal class?

3 At a youth club, the members were asked: 'How many times have you played table tennis this week?'

These are their replies.

3	6	0	13	5	0	0	2	6	8	4	14
1	0	0	0	3	0	4	7	0	1	2	1
0	14	13	0	2	0	9	13	4	5	1	0
0	11	4	3	1	0	0	0	3	4	6	10
0	1	0	0	0	0	13	2	1	3	0	0
10	6	0									

a Create a grouped frequency table:

    i with a class size of 3, i.e. 0–2, 3–5, 6–8, 9–11, …

    ii with a class size of 5, i.e. 0–4, 5–9, 10–14, ….

b What is the mode?

c Draw a bar chart for each frequency table.

d What is the modal class?

e Which class size seems more appropriate to use?

4 The table shows the times of goals scored in football matches played on one weekend in November.

Time of goals (minutes)	Frequency
$0 < \text{time} \leqslant 15$	3
$15 < \text{time} \leqslant 30$	6
$30 < \text{time} \leqslant 45$	8
$45 < \text{time} \leqslant 60$	4
$60 < \text{time} \leqslant 75$	2
$75 < \text{time} \leqslant 90$	5

a One goal was scored after exactly 75 minutes. In which class was it recorded?

b Five teams scored in the last five minutes of their games.

    Write down what other information this tells you.

c Draw a frequency diagram to represent the data in the table.

Activity: Heights in the class

A Collect data about the heights of the girls and heights of the boys in your class.

B Design frequency tables to record the information.

C Draw a frequency diagram to illustrate your table of results.

D Comment on the results.

6.6 Data collection

Learning objective

* To develop greater understanding of data collection

Suppose you wanted to organise a party to celebrate the end of term at school. You need to know what pupils would like. Here are some questions you might ask.

What shall we charge?
What time shall we start?
What time shall we finish?
What food shall we eat?

It would be difficult to ask everyone, so you could ask a sample of the pupils in your school these questions.

This means, instead of asking everyone, you ask a few from each year group.

As you ask each question, you could immediately complete your **data-collection form**.

An example of a suitable data-collection form is shown below.

Year group	Boy or girl	How much to charge?	Time to start?	Time to finish?	What would you like to eat?
Y7	B	£1	7 pm	11 pm	Crisps, beefburgers, chips
Y7	G	50p	7 pm	9 pm	Chips, crisps, lollies
Y8	G	£2	7:30 pm	10 pm	Crisps, hot dogs
Y9	B	£3	8:30 pm	11:30 pm	Chocolate, pizza
Keep track of their age.	Try to ask equal numbers of boys and girls.	Once this data is collected, it can be sorted into frequency tables.			

The five stages in running this type of data collection are:

* deciding what questions to ask and who to ask
* creating a simple, suitable data-collection form for all the questions
* asking the questions and completing the data-collection form
* after collecting all the data, collating it in frequency tables
* analysing the data to draw conclusions from the data collected.

The size of your sample will depend on many things. It may be simply the first 50 people you meet. Or you may want to target a particular fraction of the available people.

In the above example, a good sample would probably be about four from each class, two boys and two girls.

Exercise 6F

A class completed the data-collection activity described above on a sample of 10 pupils from each of years 7, 8 and 9. This is their data-collection form.

Year group	Boy or girl	How much to charge?	Time to start?	Time to finish?	What would you like to eat?
Y7	B	£1	7:00 pm	11:00 pm	Crisps, burgers, chips
Y7	G	50p	7:00 pm	9:00 pm	Chips, crisps, ice pops
Y8	G	£2	7:30 pm	10:00 pm	Crisps, kababs
Y9	B	£3	8:30 pm	11:30 pm	Chocolate, pizza
Y9	G	£2	8:00 pm	10:00 pm	Pizza
Y9	B	£2.50	7:30 pm	9:30 pm	Kababs, Chocolate
Y8	G	£1	8:00 pm	10:30 pm	Crisps
Y7	B	75p	7:00 pm	9:00 pm	Crisps, burgers
Y7	B	£1	7:30 pm	10:30 pm	Crisps, ice pops
Y8	B	£1.50	7:00 pm	9:00 pm	Crisps, chips, kababs
Y9	G	£2	8:00 pm	11:00 pm	Pizza, chocolate
Y9	G	£1.50	8:00 pm	10:30 pm	Chips, pizza
Y9	G	£2	8:00 pm	11:00 pm	Crisps, pizza
Y7	G	£1.50	7:00 pm	9:00 pm	Crisps, ice pops, chocolate
Y8	B	£2	7:30 pm	9:30 pm	Crisps, ice pops, chocolate
Y8	B	£1	8:00 pm	10:00 pm	Chips, kababs
Y9	B	£1.50	8:00 pm	11:00 pm	Pizza
Y7	B	50p	7:00 pm	9:30 pm	Crisps, kababs
Y8	G	75p	8:00 pm	10:30 pm	Crisps, chips
Y9	B	£2	7:30 pm	10:30 pm	Pizza
Y8	G	£1.50	7:30 pm	10:00 pm	Chips, kababs, chocolate
Y8	B	£1.25	7:00 pm	9:30 pm	Chips, kababs, ice pops
Y9	G	£3	7:00 pm	9:30 pm	Crisps, pizza
Y9	B	£2.50	8:00 pm	10:30 pm	Crisps, kababs
Y7	G	25p	7:30 pm	10:00 pm	Crisps, burgers, ice pops
Y7	G	50p	7:00 pm	9:00 pm	Crisps, pizza
Y7	G	£1	7:00 pm	9:30 pm	Crisps, pizza
Y8	B	£2	8:00 pm	10:00 pm	Crisps, chips, chocolate
Y8	G	£1.50	7:30 pm	9:30 pm	Chips, burgers
Y7	B	£1	7:30 pm	10:00 pm	Crisps, ice pops

1 a Create a frequency table for the suggested charges from each year group.

Charges	Tallies					
	Y7	Total	Y8	Total	Y9	Total
25p						
50p						
75p						
£1						
£1.25						
£1.50						
£2						
£2.50						
£3						

   b Comment on the differences between the year groups.

2 a Create a frequency table for the suggested starting times from each year group.

   b Comment on the differences between the year groups.

3 a Create a frequency table for the suggested finishing times from each year group.

   b Comment on the differences between the year groups.

(MR) 4 a Create and complete a frequency table as before for the food suggestions of each year group.

   b Comment on the differences between the year groups.

Investigation

Investigate the differences in the views of boys and girls about the suggested length of time for the disco.

Ready to progress?

I can use a data collection form and tally charts to collect data.
I can find the mode and range for a set of data.

I can find the median and the mean for a set of data.
I can compare two simple sets of data.
I can group data, where appropriate, into equal classes.
I know the difference between discrete and continuous data.

Review questions

1   Twenty 11-year-olds were tweeting about how much it cost them to travel to school on a bus. These are the amounts they paid.

40p	£1	75p	£1.50	£1	40p	75p	£1	£1	£1.50
50p	60p	£1.50	£1	50p	£1	£1.20	£1	75p	60p

   a   What was the modal amount spent on the bus fare?
   b   What was the mean bus fare?

2   These are the temperatures recorded on 21 February one year, in 15 major towns of the UK.

2 °C	−3 °C	1 °C	0 °C	2 °C	−1 °C	1 °C	0 °C
−3 °C	3 °C	−4 °C	−2 °C	0 °C	−2 °C	−3 °C	

   a   What was the median of the temperatures recorded?
   b   What was the mean UK temperature on 21 February?

3   Chris was asked to make a rectangle with a piece of wire of length 18 cm.

   He was told to make the length and the width whole numbers of centimetres.

   What is the range of areas he could make with that length wire?

4   When the masses of a sample of ten tins of assorted chocolates were measured, these were the results.

   2.09 kg 2.1 kg 1.95 kg 2.03 kg 1.98 kg 2.07 kg 2.02 kg 1.96 kg 1.97 kg 2.01 kg

   What was the mean mass of the ten tins of chocolates?

5   Jess was asked to draw a rectangle with the same area as a square of side 6 cm.

   She was told that the sides of the rectangle must be whole numbers of centimetres.

   a   Show that there are only four different-sized rectangles she can draw.
   b   Calculate the mean of the four different perimeters.
   c   Hence state which rectangle has the closest perimeter to the mean found.

6 Serrena asked children from two different schools: 'How do you travel to school?'

These are her results.

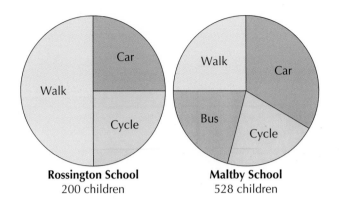

Rossington School
200 children

Maltby School
528 children

Serrena says: 'The number of children walking to Rossington school is more than the number walking to Maltby school.'

a Explain why Serrena is incorrect.

At Maltby school, one third of the children travel by car.

The number of children who cycle is the same as the number who go on the bus.

b How many of the children cycle to Maltby school?

7 Frozo asked her friends on Facebook to tell her their heights. Her friends sent her these results.

153 cm	1.55 m	1.46 m	150 cm	1.59 m	140 cm
1493 mm	1555 mm	1.54 cm	1.39 m	1452 mm	1.49 m
152 cm	159 cm	1.4 m	1418 mm	1.44 m	141 cm
1.37 m	1478 mm	139 cm	1453 mm	1.5 m	1.34 m
1549 mm	145 cm	138 cm	1.37 m	1523 mm	1.33 m

a Put the heights into a grouped frequency tally chart, using a class size of 5 cm.

b What is the modal class?

8 Draw a triangle in which the sides have a modal length of 5 cm and a mean length of 6 cm.

9 a Explain the difference between discrete and continuous data.

b Put the following data into two lists, 'discrete data' and 'continuous data'.

   time, shoe size, weight, height, colours, length, cost

Discrete data	Continuous data

Challenge

Maths tournament

Teams

1 The table shows the names and ages of the seven members of the Rowlinson School team.

   a What is the modal age?

   b What is the median age?

   c What is the mean age?

	Age (years)
Chris	13
Lewis	14
Reda	12
Anthony	14
Cheryl	13
Elizabeth	12
Jose	14

Ten question sprint

2 The boys had to answer 10 mental mathematics questions as quickly as they could.

They had to answer each question correctly before they could start the next one.

The table shows their times for the ten questions.

   a What is the modal time?

   b What is the median time?

   c What is the mean time?

	Time (seconds)
Chris	37
Lewis	40
Reda	36
Anthony	40
Jose	38

20 minute marathon

3 Before the competition, the team practised answering as many mathematical questions as they could in 20 minutes.

This graph shows how well they scored out of 150 marks during their practice sessions.

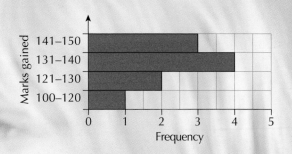

How often did they score:

a in the range 141–150   **b** more than 120   **c** less than 141?

On the buzzer

4 In this section of the tournament, the quizmaster fired mental mathematics questions at the teams.

Any contestant could press the buzzer and answer.

This is a tally of all the times each contestant from two schools answered correctly.

a What was the modal number of correct answers from all contestants?

b What was the mean number of correct answers for each team?

c Which team won this part of the competition?

d Which contestant answered correctly most frequently?

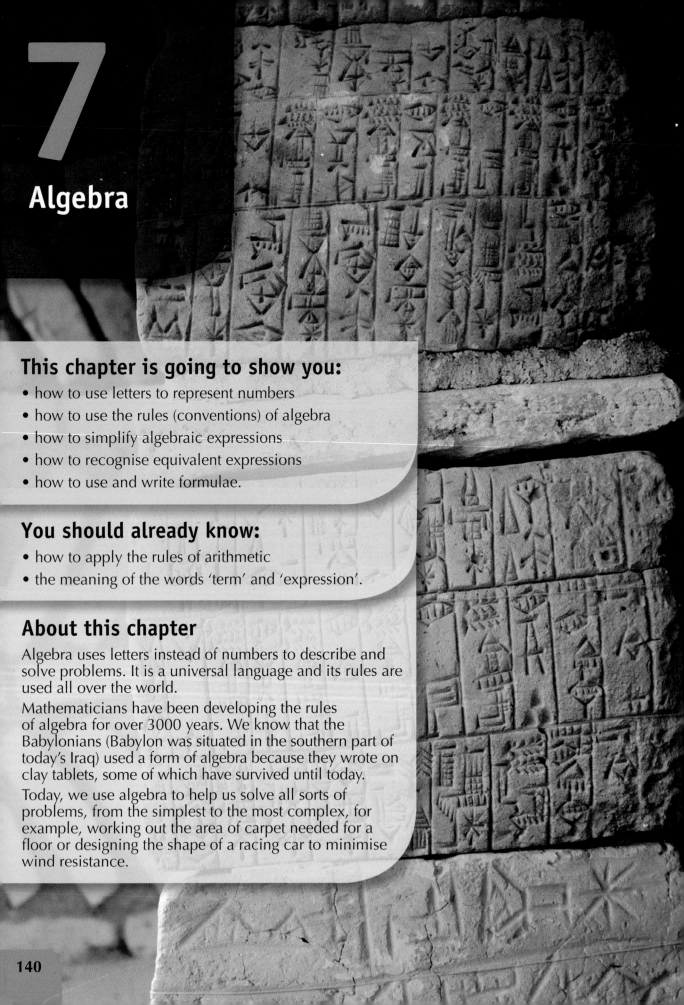

7

Algebra

This chapter is going to show you:

- how to use letters to represent numbers
- how to use the rules (conventions) of algebra
- how to simplify algebraic expressions
- how to recognise equivalent expressions
- how to use and write formulae.

You should already know:

- how to apply the rules of arithmetic
- the meaning of the words 'term' and 'expression'.

About this chapter

Algebra uses letters instead of numbers to describe and solve problems. It is a universal language and its rules are used all over the world.

Mathematicians have been developing the rules of algebra for over 3000 years. We know that the Babylonians (Babylon was situated in the southern part of today's Iraq) used a form of algebra because they wrote on clay tablets, some of which have survived until today.

Today, we use algebra to help us solve all sorts of problems, from the simplest to the most complex, for example, working out the area of carpet needed for a floor or designing the shape of a racing car to minimise wind resistance.

7.1 Expressions and substitution

Learning objectives

- To use algebra to write simple expressions and recognise equivalent expressions
- To substitute numbers into expressions to work out their value

Key words

equivalent	expression
substitute	term
variable	

Look at this rectangular business card.

The card is 5.5 cm wide. You do not know the length. It is labelled as a cm. You can use letters or symbols like this to represent unknown values.

The perimeter of a rectangle is the distance round the outside. For this rectangle you can write that as $a + 5.5 + a + 5.5$ cm or more simply as $2a + 11$ cm.

a cm

FRJ

Frank R Johnson
14 The Square
Briteon   05748 256458

5.5 cm

This is an **expression** for the perimeter of this rectangle.

$2a$ and 11 are called the **terms** of the expression $2a + 11$. Notice that you leave out the $\times$ sign when you write terms and expressions.

a is called a **variable** because it can be given different values.

Another way to find the perimeter is to add the length and width and double the answer. An expression for that is $2(a + 5.5)$ cm.

Now suppose the length is 9 cm. That means that $a = 9$. You can **substitute** this value into both expressions.

Then   the value of $2a + 11 = 2 \times 9 + 11 = 29$

and   the value of $2(a + 5.5) = 2 \times (9 + 5.5) = 2 \times 14.5 = 29$.

Both expressions give the same value for the perimeter of the rectangle, which is 29 cm. They are **equivalent** expressions.

To find the area of a rectangle you multiply width by length.

An expression for the area of this business card is $5.5a$ cm².

Example 1

Alice, on the right, is x years old.

Bella, on the left, is seven years younger than Alice.

Cheryl, in the middle, is four times as old as Bella.

Write an expression for:

a Bella's age   **b** Cheryl's age.

  a Subtract 7 from Alice's age to get Bella's age.   The expression is $x - 7$.

  b Multiply Bella's age by 4 to find Cheryl's age.   The expression is $4(x - 7)$.

Example 2

Work out the value of each expression, if $p = 3.6$ and $q = 6.5$.

a $4(p - 1)$ **b** $3q + 8.2$ **c** q^2 **d** $\frac{p}{4}$ **e** $3(p + q)$

  a If $p = 3.6$ then $4(p - 1) = 4 \times (3.6 - 1) = 4 \times 2.6 = 10.4$

  b If $q = 6.5$ then $3q + 8.2 = 3 \times 6.5 + 8.2 = 19.5 + 8.2 = 27.7$

  c q^2 is 'q squared', which is $q \times q = 6.5 \times 6.5 = 42.25$

  d $\frac{p}{4}$ means $p \div 4$, which is $3.6 \div 4 = 0.9$

  e $3(p + q) = 3 \times (3.6 + 6.5) = 3 \times 10.1 = 30.3$

Exercise 7A

1   Mike is three years older than his wife Karen.

Karen is five times as old as her son George.

Suppose George is g years old.

  a   Write an expression, containing g, for Karen's age.

  b   Write an expression for Mike's age.

  c   Work out the ages of the three people, when $g = 5$.

  d   Work out the ages of the three people, when $g = 7$.

2   Kate's mass is k kilograms.

Write down an expression for the mass, in kilograms, of each of these children.

  a   Jason's mass is 3.5 kg more than Kate's.

  b   Keira's mass is half Kate's mass.

  c   Becky's mass is 1.5 kg less than Kate's.

  d   Mark's mass is twice the mass of Kate.

  e   Andy's mass is twice the mass of Jason.

  f   Lara's mass is 1 kg more than Keira's.

3   Write down an expression for the perimeter, in centimetres, of each shape.

a
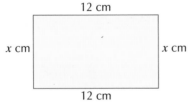

b
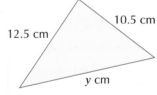

c
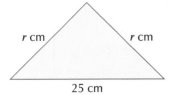

d
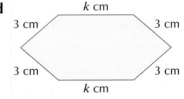

4 **a** Look again at your answer to question **3a**.

   Work out the value of your expression, when $x = 7.5$.

   b Look again at your answer to **3c**.

   Work out the value of your expression, when $r = 18$.

   c Look again at your answer to **3d**.

   Work out the value of your expression, when $k = 5$.

5 The perimeter of this triangle is p cm.

   a Work out an expression for the length of the missing side.

   b Work out the length of the missing side, when $p = 39$.

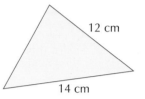

6 The perimeter of this equilateral triangle is d cm.

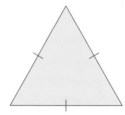

   Write down an expression for the length, in centimetres, of each side.

PS **7** The perimeter of this isosceles triangle is m cm.

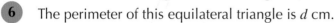

   The length of the longest side is 30 cm.

   Write down an expression for the length, in centimetres, of each of the other two sides.

8 Work out the value of each of these expressions, when $t = 9$.

   a $2t - 5$ **b** $2(t - 5)$ **c** $t^2 + 1$ **d** $(t + 1)^2$ **e** $\dfrac{t}{4.5}$

9 Work out the value of each of these expressions when $r = 5$ and $s = 6$.

   a $3r + s$ **b** $3(r + s)$ **c** $r + 3s$ **d** $3r - s$ **e** $\dfrac{r + s}{2}$

10 **a** Matt thinks of a number.

   He doubles it. Then he adds 4.

   Suppose the number he thinks of is n.

   Work out an expression for his final answer.

   b Lucy thinks of a number.

   She subtracts 7 and then multiplies the answer by 5.

   Suppose the number she thinks of is t.

   Work out an expression for her final answer.

a The perimeter of the square in the diagram is p cm.

Show that the perimeter of the rectangle is $1.5p$.

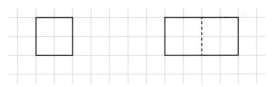

b Work out an expression, in terms of p, for the perimeter of this rectangle.

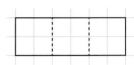

c Work out an expression, in terms of p, for the perimeter of this square.

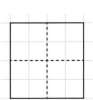

MR 12 a Asif says: 'The perimeter of this shape is $4a + 20$ cm.'

Show that this expression is correct.

b Work out the value of $4a + 20$, if $a = 6$.

c Sharan says: 'The perimeter of this shape is $2(2a + 10)$ cm.'

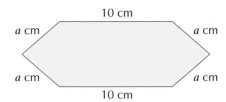

Show that this formula gives the same value as the formula in part **b**, if $a = 6$.

d Use the diagram to explain why the formula in part **c** is correct.

e Tariq says: 'The perimeter of this shape is $4(a + 5)$ cm.'

Show that this formula gives the same value as those in parts **b** and **c**, if $a = 6$.

f Use the diagram to explain why the formula in part **c** is correct.

PS 13 The perimeter of the square is p cm and the perimeter of the circle is q cm.

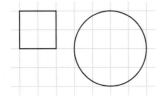

Work out an expression for the perimeter of each of these shapes.

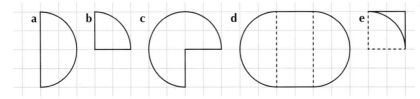

PS 14 Look again at the shapes in question **13**.

The area of the square is s cm^2 and the area of the circle is c cm^2.

Work out an expression for the area of each of the other shapes.

Investigation: Areas of compound shapes

The areas of these two shapes are C cm² and D cm².

A Draw different compound shapes made up from copies of these two basic shapes.

B Write an expression for the area of each one.

C The lengths of the sides of the rectangle are e cm and $2e$ cm. The perimeter of the semicircle is f cm.

Try to write an expression for the perimeter of each of the shapes you drew in part **A**.

C cm² D cm²

7.2 Simplifying expressions

Learning objective

• To learn how to simplify expressions

Key words	
coefficient	like terms
simplify	unlike terms

This diagram shows the dimensions, in metres, of the floor of a room.

You can find the perimeter of the room by adding all the lengths.

In metres, this is:

$b + 3 + 2a + b + a + 3 + a$

You can simplify this by adding the a terms, the b terms and the numbers separately.

$b + b = 2b \quad 2a + a + a = 4a \quad 3 + 3 = 6$

The perimeter is $2b + 4a + 6$ metres. You cannot combine these terms any more.

$2a$ and a are both multiples of a. They are called **like terms** and they can be added together.

$4a$ and 6 are **unlike terms** and they cannot be added together.

In the expression $2b + 4a + 6$, the number 2 is the **coefficient** of b and 4 is the coefficient of a.

Simplifying an expression in this way is called 'collecting like terms'.

Example 3

This is the floor plan of a room. The lengths are in metres.

Work out an expression for the area, in square metres. Write it as simply as possible.

The area of the floor is the sum of the areas of the two rectangles.

The area is $(b \times a) + (a \times 2b)$.

You can write $b \times a$ as ba or as ab. You do not need a × sign.

$a \times 2b$ means $a \times 2 \times b$ and you can write that as $2ab$.

The area, in square metres, is $ab + 2ab = 3ab$.

Example 4

Simplify each expression.

a $5x - 3y - 2x + y - x$ **b** $4d - 5 - 3d$ **c** $ef + 4e + 2f$

  a Look at the x-terms. $5x - 2x - x = 2x$

       Look at the y-terms. $-3y + y = -2y$

       The expression simplifies to $2x - 2y$.

  b $4d - 3d = d$

       The expression simplifies to $d - 5$.

  c There are no like terms in this expression. It cannot be simplified.

Exercise 7B

1 Write each expression as a single term.

  a $k + 2k$ **b** $5b + 3b$ **c** $6c - 4c$ **d** $4d + d + 4d$

  e $x + 2x - x$ **f** $8y - 7y$ **g** $z + 5z - 3z$ **h** $2p + 3p - p + 2p$

2 Simplify each expression as much as possible.

  a $2f + 3 + 3f + 4$ **b** $5t - 3 + t - 2$ **c** $2x + y + x + 3y$ **d** $5g - h - 2h$

  e $2w + 2w + 5 + 5$ **f** $4 + 6y - 2 - 2y$ **g** $r - 2s + 2r + s$ **h** $4 + 3t - 3 - 2t$

3 The lengths of the sides of these shapes are given in centimetres.

Work out an expression for the perimeter of each shape. Write your answers as simply as possible.

a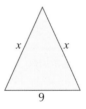

b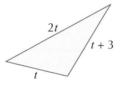

c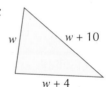

d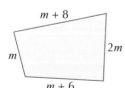

e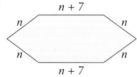

4 The length of each side of this square is a cm.

The area of the square is a^2 cm^2.

Work out an expression for the area of each of these shapes.

a

b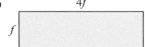

c

5 This is a number wall.

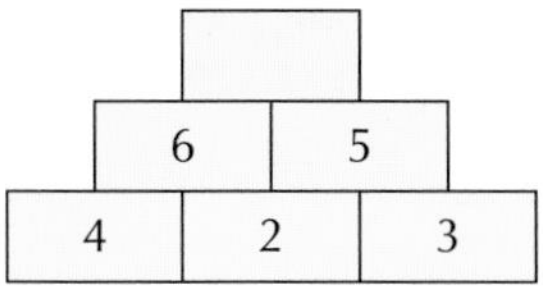
6
5
4
2
3

Add the numbers in the two bricks below, to get the number in the brick above.

$4 + 2 = 6$ and $2 + 3 = 5$

The missing top number is 11.

Work out an expression for the top brick in each of these number walls.

Write the expressions as simply as possible.

a

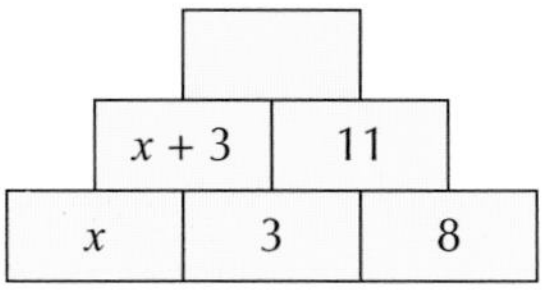
x + 3
11
x
3
8

b

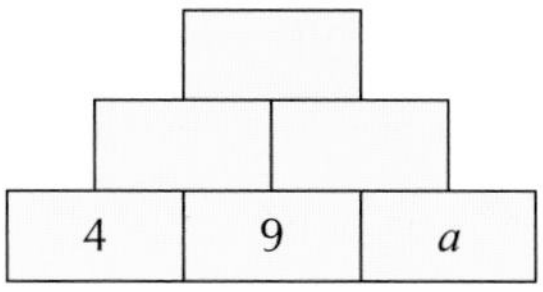
4
9
a

c

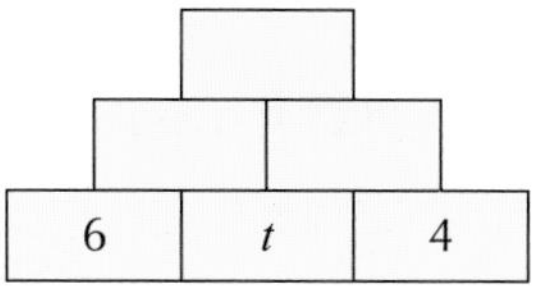
6
t
4

d

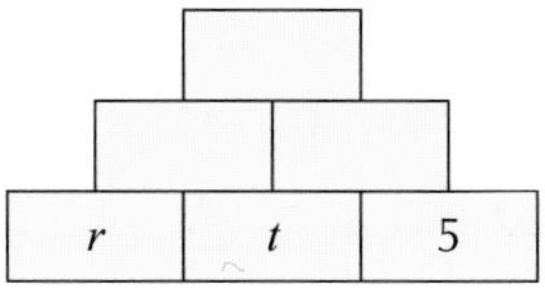
r
t
5

e

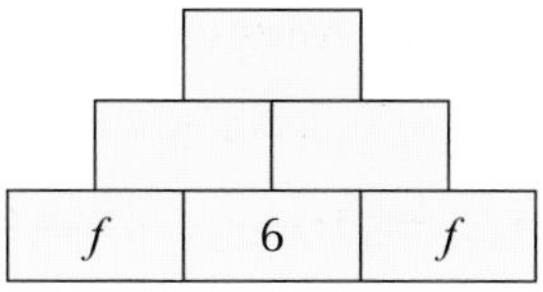
f
6
f

f

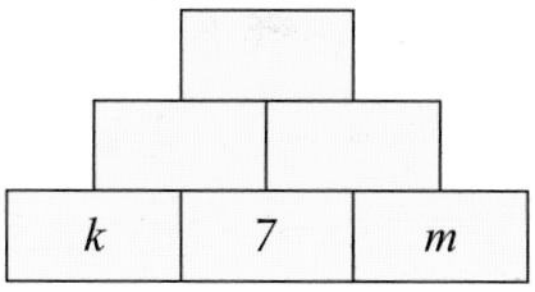
k
7
m

6 The lengths of the sides of these shapes are given in centimetres.
Work out an expression for the perimeter of each shape.

Write your answers as simply as possible.

a

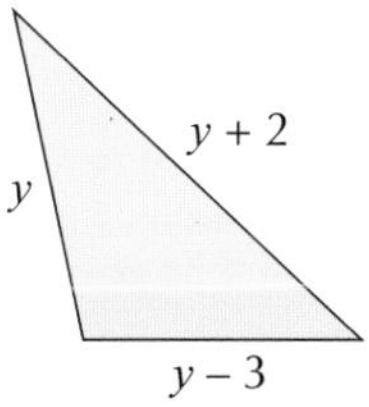
y + 2
y
y – 3

b

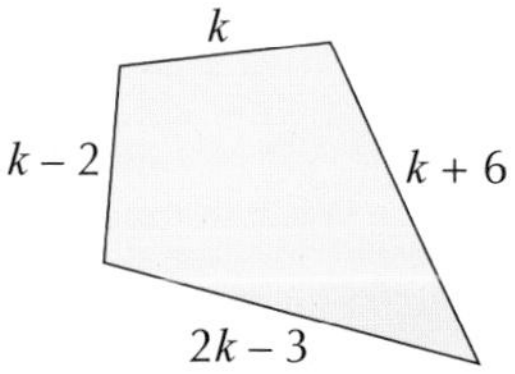
k
k – 2
k + 6
2k – 3

c

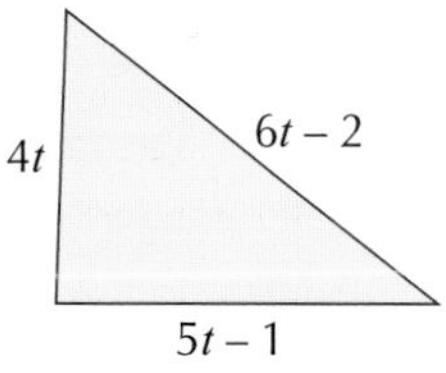
4t
6t – 2
5t – 1

7 Work out an expression for the perimeter of each shape.
Write your answers as simply as possible.

a

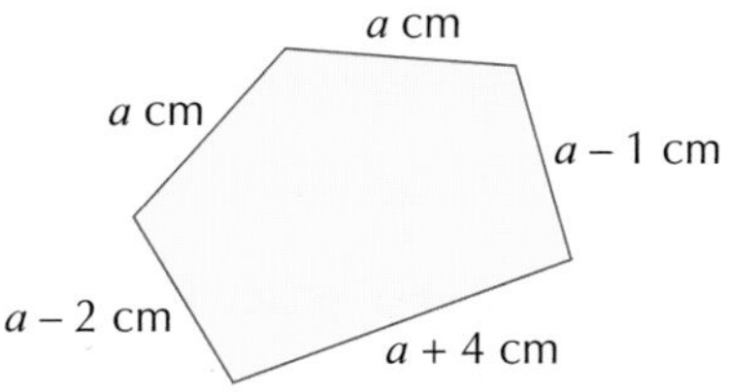
a cm
a cm
a – 1 cm
a – 2 cm
a + 4 cm

b

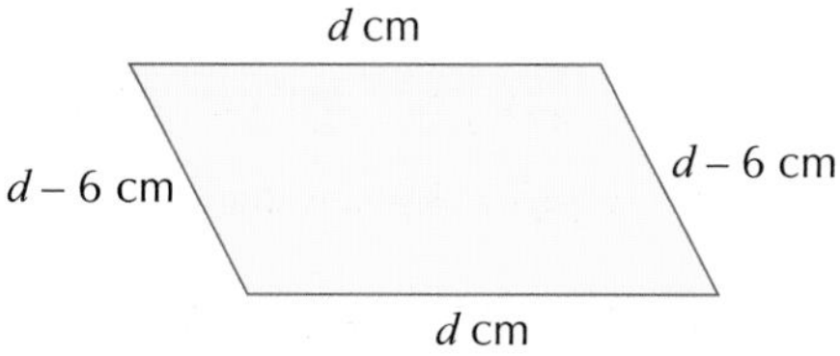
d cm
d – 6 cm
d – 6 cm
d cm

PS **8** Work out an expression for the area of the region shaded green in this diagram.

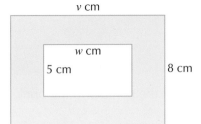

PS **9** Find an expression for the area of each shape.

a

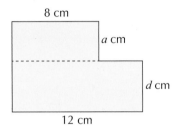

8 cm

a cm

d cm

12 cm

b

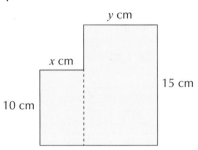

y cm

x cm

10 cm

15 cm

c

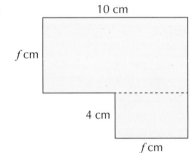

10 cm

f cm

4 cm

f cm

d

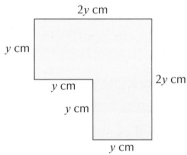

2*y* cm

y cm

y cm

2*y* cm

y cm

y cm

MR **10** For these number walls, write an expression for each of the empty bricks.

a

	a	
15		
6	9	

b

	k	
		13
	6	7

Investigation: Adding expressions

Here are 10 expression cards.

$x + 1$ $x - 1$ $2x + 1$ $2x + 3$ $2x$

$x + 2$ $x - 2$ $3x + 4$ $3x - 1$ $2x - 1$

The expressions on the cards can be added together, like this.

$x + 1$ + $x + 2$ = $2x + 3$

This example shows that the expression on the $2x + 3$ card can be written as the sum of the expressions on two of the other cards.

Which cards can be written as the sum of the expressions on two or more of the other cards?

7.3 Using formulae

Learning objective

* To use formulae

Key words

| formula | formulae |

You have already used simple **formulae** in Chapter 3.

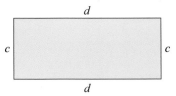

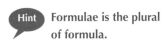

> **Hint** Formulae is the plural of formula.

The area, A, of this rectangle is given by the formula:

$$A = cd$$

The perimeter, P, of this rectangle is given by the formula:

$$P = 2c + 2d \text{ or } P = 2(c + d)$$

The volume (V) of a cube or cuboid is equal to length (l) × width (w) × height (h).

$$V = l \times w \times h \text{ or } V = lwh$$

Example 5

The monthly cost, £c, of a phone contract is given by the formula $c = 15 + 0.05n$, where n is the number of calls. Work out the monthly cost, if 83 calls are made.

83 calls are made so $n = 83$.

$$c = 15 + 0.05n$$
$$= 15 + 0.05 \times 83$$
$$= 15 + 4.15 \quad \text{Do the multiplication first.}$$
$$= 19.15$$

The cost is £19.15.

Example 6

The braking distance for a car is the distance it travels to come to a halt, after the brakes have been applied. The braking distance, d metres, for a car travelling at v mph is given by the formula $d = \dfrac{v^2}{65}$.

Work out the braking distance for a car travelling at 30 mph.

In this case $v = 30$.

$$d = \frac{v^2}{65}$$
$$= \frac{30^2}{65} \qquad 30^2 = 30 \times 30 = 900$$
$$= \frac{900}{65} \qquad \text{This means } 900 \div 65.$$
$$= 14 \text{ to the nearest whole number}$$

The breaking distance is 14 m.

Exercise 7C

1 A plumber uses this formula to find the cost (£c) when he is called out for an emergency:

$$c = 30h + 45$$

where h is the number of hours the job takes.

Work out the cost if the job takes:

a 2 hours **b** 4 hours **c** 30 minutes.

2 The cost (£h) of hiring a van for a day is given by the formula:

$$h = 30 + 0.09m$$

where m is the number of miles travelled.

Work out the cost, if the van is driven:

a 50 miles **b** 100 miles **c** 217 miles.

3 In rugby union, there are three ways to score points: a try, a conversion or a goal.

If a team scores t tries, c conversions and g goals, the total number of points, p, is given by the formula:

$$p = 5t + 2c + 3g$$

Work out the total number of points scored for three tries, two conversions and four goals.

4 Gas used in a home is measured in units.

If a family uses u units in a three-month period, the cost, £c, is given by the formula:

$$c = 9.2 + 0.8u$$

Work out the cost of using:

a 100 units **b** 150 units **c** 200 units

5 In this equilateral triangle, the length of each side is x cm.

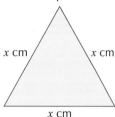

x cm x cm

x cm

A formula for the area, a cm², of this triangle is $a = 0.43x^2$.

a Work out the area of an equilateral triangle with a side of length 3 cm.

b Copy and complete this table.

x	5	10	20
a	10.75		

6 An athlete runs d metres in t seconds.

The athlete's average speed, s metres per second, is given by the formula $s = \dfrac{d}{t}$.

Work out the average speed of an athlete who runs:

a 400 m in 50 seconds

b 1.5 km in 4 minutes.

7 This can is in the shape of a cylinder.

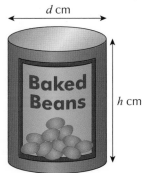

A formula for the area, a cm², of the curved surface of the can is $a = 3.14dh$.

Work out the area of the curved surface of a can if $d = 8$ and $h = 11$. Give your answer to the nearest whole number.

 8 This is an isosceles triangle.

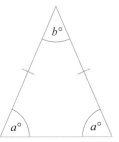

A formula for the size of each of the bottom angles is $a = \dfrac{180 - b}{2}$.

a Work out the value of a if $b = 20$.

b Copy and complete this table showing possible values of b and a.

b	20	30	40	50	60
a					

c Show that an equivalent expression for the bottom angle is $90 - \dfrac{b}{2}$.

The diagram shows a regular five-sided polygon and a regular six-sided polygon.

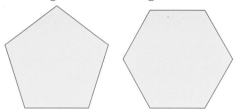

All the angles of a regular polygon are the same size.

If a regular polygon has n sides and the size of the angle of the polygon is $a°$, then the value of a is given by the formula:

$$a = 180 - \frac{360}{n}$$

a Copy and complete this table to show the sizes of the angles for some regular polygons.

Number of sides, n	5	6	8	10	12	20
Angle in degrees, a						

b Show that the expression $180\left(1 - \frac{2}{n}\right)$ gives the same answers.

Challenge: Changing temperatures

Temperature can be measured in Celsius (C) or Fahrenheit (F).

A A formula to convert from Celsius to Fahrenheit is $F = 1.8C + 32$.

Copy and complete this table.

C	0	20	40	60
F				

B A formula to convert from degrees Fahrenheit to degrees Celsius is:

$$C = \frac{F - 32}{1.8}$$

Copy and complete this table.

F	32	50	86	122	212
C					

C Scientists measure temperature in kelvins (K).

A formula to convert from degrees Celsius to kelvins is $K = C + 273$.

a Convert 35 °C to kelvins.

b Work out a formula to convert from Fahrenheit to kelvins.

7.4 Writing formulae

Learning objective

- To write formulae

Sometimes you will need to create your own formula to represent a statement or to solve a problem.

Example 7

The monthly cost of a phone contract is £8.50 plus 5p per text.

Work out a formula for the cost, £c, if n texts are sent in a month.

The cost, in pounds, of n texts is $0.05n$.	5p is £0.05.
That means $c = 0.05n + 8.5$	You must add £8.50 to the cost of the texts.

The formula could also be written as $c = 8.5 + 0.05n$.

Example 8

The length of a rectangle is l cm. The width is 15 cm less than the length. Find a formula for:

a the area, a cm²   **b** the perimeter, p cm.

a The width is $l - 15$ cm.

The area is the length × the width so $a = l(l - 15)$ The brackets are important here.

b The perimeter is 2 × (length + width) so $p = 2(l + l - 15)$

which simplifies to $p = 2(2l - 15)$

Example 9

Angie thinks of a number. She doubles it. She takes the answer away from 100. Then she divides by 4. If the initial number is n, write a formula for the final answer, f.

She doubles n to get $2n$.

She takes this away from 100.   That gives $100 - 2n$.

Now divide by 4 to get: $f = \dfrac{100 - 2n}{4}$

Exercise 7D

1 The cost of a mobile phone contract is in two parts.

There is a fixed charge of £14 per month plus 10p for every phone call.

a Work out a formula for the total cost, £t, if there are c calls.

b Work out the total cost if there are 137 calls.

c The fixed charge is increased to £17 and the cost of each call is reduced to 8p.

Write down the new formula for the cost.

2 A restaurant has a special menu for group bookings. There is a service charge of £10 plus £16 per person for the meal.

Write down a formula for the total cost, £c, for a party of n people.

3 A recipe books give these instructions for working out the time to roast a chicken.

Cook for 40 minutes per kilogram plus an extra 15 minutes.

a Write down a formula for the cooking time, t minutes, for a w kg chicken.

b Alter the formula to give the cooking time in hours.

4 Oranges cost 40p each. Bananas cost £1.70 per kilogram.

Work out a formula for the cost, £c, of r oranges and t kg of bananas.

5 Darren is thinking of a number.

He adds 12 to his number and then multiplies the result by 4.

Suppose Darren's initial number is d.

Write down a formula for his final result, f.

6 Here are some more 'Think of a number' questions, similar to question 5.

In each case, write a formula for the final answer, t, in terms of the starting number, n.

a Think of a number, multiply by 5, then subtract 7.

b Think of a number, halve it, then add 10.

c Think of a number, multiply by 4, subtract 3, then divide by 2.

d Think of a number, multiply by 5, then take the result away from 200.

e Think of a number, double it, add 3, then double it again.

(PS) 7 The width of a rectangle is w cm.

The length is 3 cm more than the width.

a Write down a formula for the area, a cm², in terms of w.

b Write down a formula for the perimeter, p cm, in terms of w.

(MR) 8 Consecutive whole numbers are whole numbers that 'follow on' in sequence, such as 12, 13, 14 and 15.

N is the smallest of four consecutive whole numbers.

a Write down expressions for the other three numbers.

b The four consecutive numbers, starting with N, are the lengths of the sides of a quadrilateral. Work out a formula for the perimeter, P, of the quadrilateral. Write the formula as simply as possible.

9 To find the average number of goals scored in three football matches, you find the total number of goals scored and divide by three.

The numbers of goals scored in three matches are a, b and c.

Write down a formula for the average, A.

10 Here is a number wall.

The number in each brick is the sum of the two numbers below it.

Write a formula for c in terms of a and b.

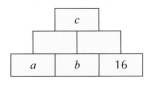

11 Here is a number wall.

Write a formula for c in terms of d.

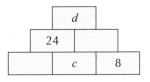

MR **12** Here is a number wall with four rows of bricks.

Work out a formula for t in terms of a and b.

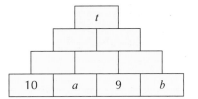

MR **13** Here is a number wall.

Work out a formula for x in terms of m.

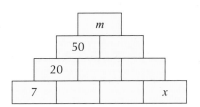

Investigation: Dotty shapes

You will need centimetre-square dotted paper for this investigation.

Here are three shapes.

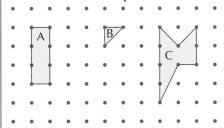

A Copy this table and complete the first three lines.

Shape	Number of dots on the perimeter (n)	Area (a cm²)
A		
B		
C		

B Draw some more shapes on centimetre-square dotted paper.

There must not be any dots inside the shape.

For example, this shape ⬠ has a dot inside so it is not allowed.

C Add your shapes to the table.

D Work out a formula for a in terms of n.

Ready to progress?

Review questions

1   Write down an expression for the perimeter of each of these shapes. Write each answer as simply as possible.

a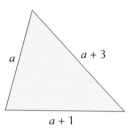

b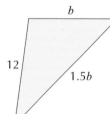

c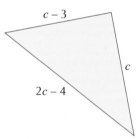

2   The area of this square is Q cm² and the area of the semicircle is S cm².   This shape has an area of $Q + 2S$ cm².

a   Write the area of each shape, in terms of Q and S.

i 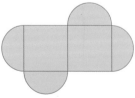   **ii**   **iii**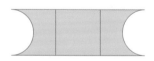

b   Draw diagrams to show shapes with these areas.
   i $2Q + 6S$ cm²   **ii** $2(Q + S)$ cm²   **iii** $2(Q - S)$ cm²

3   Work out the value of each expression, when $r = 2.5$ and $t = 6$.

   a $4(t - r)$ **b** $4t - r$ **c** $2(3t - 10)$ **d** $(t - 2)^2$

   e $\frac{t}{3} + 4r - 3$ **f** $\frac{6r + t}{3}$

(PS) **4** All the lengths in this shape are given in centimetres.

    a Work out an expression for the perimeter of the shape.

    b Work out an expression for the area of the shape.

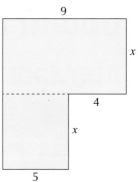

5 This shape is made from three regular pentagons.
The length of each edge is x cm.

    a The perimeter of the shape is p cm. Write down a formula for p in terms of x.

    b A formula for the area, A, of a pentagon with a side of length x is $A = 1.72x^2$.

        i Write down a formula for the area of the shape.

        ii Work out the area of the shape if each edge is 10 cm.

6 The lengths of the sides of this rectangle are as shown.

    a Work out a formula for the perimeter, P. Write it as simply as possible.

    b Write down a formula for the area, A.

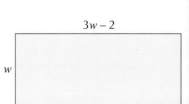

7 Kerry thinks of a number. She doubles it, adds 10 and then divides the answer by 3.

Write a formula for the result, r, in terms of the initial number, n.

8 Four numbers are n, $n + 4$, $2n$ and $2n - 7$.

    a Given that $n = 8.5$ work out:

        i the four numbers   **ii** the mean of the four numbers.

    b Work out a formula for the mean, m, of the four numbers. Write it as simply as possible.

(MR) **9** Here is a number wall.

    a Work out an expression for the number in the top brick.

    b Show that the number in the top brick can be written as $3(3x - 1)$.

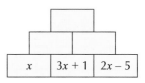

(PS) **10** These are the first five terms of a sequence.

        $8 \quad 8 + d \quad 8 + 2d \quad 8 + 3d \quad 8 + 4d$

    a Write down the first five terms of the sequence, given that $d = 3$.

    b Write down the first five terms of the sequence, given that $d = 10$.

    c Write down an expression for the sixth term.

    d Write down an expression for the tenth term.

    e Write down an expression for the 100th term.

Problem solving
Winter sports

You can hire equipment for skiing and snowboarding.

The cost depends on the number of days for which you want to hire the equipment.

A hire company uses these formulae to work out the cost.

	Skiing	Snowboarding
Adults	$C = 8 + 17D$	$C = 9 + 23D$
Children	$C = 4 + 9D$	$C = 6 + 12D$

C is the cost in pounds.
D is the number of days of hire.

1 Work out the cost of skiing for one adult for eight days.

2 Work out the total cost for two adults and three children for skiing for six days.

3 Work out the difference in cost between an adult and a child for skiing for:

   a three days   **b** five days   **c** ten days.

4 Work out a formula for the difference in cost between an adult and a child for skiing for D days.

5 Work out a formula for the difference in cost between an adult and a child for snowboarding for D days.

6 A group of three adults and twelve children are hiring equipment for four days. Two adults and half the children are skiing, the rest are snowboarding. Work out the total cost for this group.

7 a Draw axes like this on graph paper. Plot crosses to show the cost of ski hire for one adult for from one to ten days. One cross has been plotted for you.

b Draw a dashed line through your ten points. It should be a straight line.

c On the same graph, plot and join crosses to show the cost of ski hire for one child.

8 Copy the graph from question **7**.
Draw lines to show the cost of snowboard hire for an adult and for a child. Label your lines clearly.

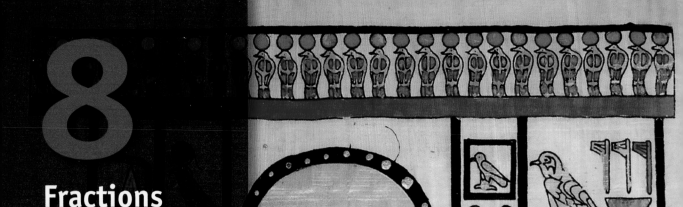

8

Fractions

This chapter is going to show you:

- how to find equivalent fractions
- how to write a fraction in its simplest form
- how to add and subtract fractions with different denominators
- how to convert a simple improper fraction to a mixed number
- how to convert a mixed number into an improper fraction
- how to add and subtract mixed numbers with different denominators.

You should already know:

- how to recognise and use simple fractions
- how to compare and order fractions with the same denominator.
- how to find the common multiples of two numbers
- how to add and subtract fractions and mixed numbers with the same denominator.

About this chapter

For thousands of years the fraction system used in Europe was based on the ancient Egyptian one. Egyptian fractions were all unitary fractions, that means they only had 1 as a numerator, for example, $\frac{1}{2}, \frac{1}{4}, \frac{1}{6}$. Big fractions were written as unitary fractions added together, for example, $\frac{3}{4}$ would be $\frac{1}{2} + \frac{1}{4}$.

The fractions $\frac{1}{2}, \frac{1}{4}, \frac{1}{8}, \frac{1}{16}$ and $\frac{1}{64}$ were sacred. They were linked to the Eye of the Horus, an Egyptian god, and were all used as measures for grain. You will have a chance to try Egyptian fractions in this chapter!

8.1 Equivalent fractions

Learning objectives

- To find equivalent fractions
- To write fractions in their simplest form

In the fraction $\frac{3}{4}$, the number on the top, 3, is the **numerator** and the number on the bottom, 4, is the **denominator**.

The term **equivalent** means 'of equal value'. You can find **equivalent fractions** by multiplying the numerator and the denominator of a fraction by the same number.

For example, if you multiply the numerator and the denominator of $\frac{3}{4}$ by 5 you get $\frac{3}{4} = \frac{15}{20}$.

You can illustrate this with a diagram.

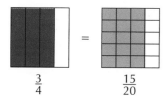

$$\frac{3}{4} \qquad \frac{15}{20}$$

Example 1

Fill in the missing number in each of these equivalent fractions.

a $\frac{2}{3} = \frac{a}{15}$ **b** $\frac{5}{8} = \frac{30}{b}$ **c** $\frac{40}{48} = \frac{c}{12}$

   a $3 \times ? = 15$ The missing number is 5.

     Multiply the numerator by 5 to get:

     $\frac{2}{3} = \frac{10}{15}$ $a = 10$

   b $5 \times ? = 30$ The missing number is 6.

     Multiply the denominator by 6 to get:

     $\frac{5}{8} = \frac{30}{48}$ $b = 48$

   c $48 \div ? = 12$ In this case divide, because the denominator is smaller.

                     The missing number is 4.

     Divide the numerator by 4 to find the value of c:

     $\frac{40}{48} = \frac{10}{12}$ $c = 10$

A fraction is in its **simplest form** if you cannot divide its numerator and denominator by the same whole number, other than 1.

If you are asked to **simplify** a fraction it means finding its simplest form.

Example 2

Write each of these fractions in its simplest form.

a $\dfrac{18}{24}$ **b** $\dfrac{27}{45}$

> **a** 2 is a **common factor** of 18 and 24, that is, it will divide into both numbers.
>
> Divide top and bottom by 2.
>
> $\dfrac{18}{24} = \dfrac{9}{12}$ This is not in the simplest form.
>
> 3 is a common factor of 9 and 12.
>
> Divide top and bottom by 3.
>
> $\dfrac{9}{12} = \dfrac{3}{4}$ This is the simplest form.
>
> You could reach the answer in one step by noticing that 6 is a common factor of 18 and 24 and dividing by 6.
>
> **b** 3 and 9 are common factors of 27 and 45.
>
> Divide top and bottom by 9.
>
> $\dfrac{27}{45} = \dfrac{3}{5}$ This is the simplest form.

Exercise 8A

  1 Here is part of a multiplication table.

7	14	21	28	35	42
8	16	24	32	40	48
9	18	27	36	45	54

Use the rows in the table to write fractions that are equivalent to these fractions.

a $\dfrac{7}{9}$ **b** $\dfrac{8}{9}$

 2 Look at the fractions listed here.

$\dfrac{5}{15}$ $\dfrac{8}{12}$ $\left(\dfrac{45}{60}\right)$ $\dfrac{14}{20}$ $\dfrac{18}{24}$ $\dfrac{30}{45}$

$\dfrac{14}{42}$ $\dfrac{49}{70}$ $\dfrac{19}{57}$ $\dfrac{75}{100}$ $\dfrac{36}{54}$ $\dfrac{42}{60}$

Write down all the fractions that are equivalent to each of these fractions.

a $\dfrac{1}{3}$ **b** $\dfrac{2}{3}$ **c** $\dfrac{3}{4}$ **d** $\dfrac{7}{10}$

3 Fill in the missing number in each pair of equivalent fractions.

a $\dfrac{2}{3} = \dfrac{\square}{36}$ b $\dfrac{3}{8} = \dfrac{\square}{24}$ c $\dfrac{5}{9} = \dfrac{\square}{81}$

d $\dfrac{2}{5} = \dfrac{\square}{35}$ e $\dfrac{3}{7} = \dfrac{\square}{56}$ f $\dfrac{4}{9} = \dfrac{\square}{72}$

g $\dfrac{1}{5} = \dfrac{\square}{75}$ h $\dfrac{2}{11} = \dfrac{28}{\square}$ i $\dfrac{4}{9} = \dfrac{40}{\square}$

4 Write each fraction in its simplest form.

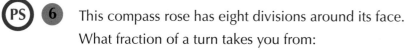

a $\dfrac{8}{12}$ b $\dfrac{12}{18}$ c $\dfrac{28}{42}$ d $\dfrac{45}{60}$ e $\dfrac{36}{40}$ f $\dfrac{200}{500}$

g $\dfrac{16}{24}$ h $\dfrac{42}{84}$ i $\dfrac{20}{24}$ j $\dfrac{16}{20}$ k $\dfrac{32}{48}$ l $\dfrac{75}{100}$

m $\dfrac{12}{28}$ n $\dfrac{24}{18}$ o $\dfrac{180}{270}$ p $\dfrac{400}{200}$ q $\dfrac{128}{120}$ r $\dfrac{240}{40}$

(PS) 5 There are 23 fractions in the sequence: $\dfrac{1}{24}, \dfrac{2}{24}, \dfrac{3}{24}, \dfrac{4}{24}, \dots, \dfrac{23}{24}$.

How many of them cannot be simplified?

(PS) 6 This compass rose has eight divisions around its face.

What fraction of a turn takes you from:

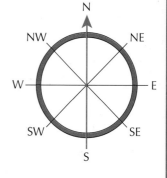

a NW to SE   b E to SW   c NE to SW
   clockwise   anticlockwise   clockwise

d SW to NE   e W to SW   f NE to NW
   anticlockwise   clockwise   clockwise?

Write your fractions in their simplest form.

7 Here are two fractions.

$\dfrac{3}{4}$ $\dfrac{2}{3}$

a Write fractions equivalent to $\dfrac{3}{4}$ and $\dfrac{2}{3}$ that have the same numerator.

b Write fractions equivalent to $\dfrac{3}{4}$ and $\dfrac{2}{3}$ that have the same denominator.

8 Here are two fractions.

$\dfrac{5}{8}$ $\dfrac{7}{10}$

a Write fractions equivalent to $\dfrac{5}{8}$ and $\dfrac{7}{10}$ that have the same numerator.

b Write fractions equivalent to $\dfrac{5}{8}$ and $\dfrac{7}{10}$ that have the same denominator.

9 a One metre is 100 cm. What fraction of a metre is 35 cm?

b One kilogram is 1000 grams. What fraction of a kilogram is 550 grams?

c One hour is 60 minutes. What fraction of 1 hour is 33 minutes?

d One kilometre is 1000 metres. What fraction of a kilometre is 75 metres?

Give each answer in its simplest form.

PS **10** Work out these quantities, giving each answer in its simplest form.

    a Calculate the fraction of a metre that is equivalent to:

       i 85 cm   **ii** 250 mm   **iii** 40 cm.

    b Calculate the fraction of a kilogram that is equivalent to:

       i 450 g   **ii** 45 g   **iii** 975 g.

    c Calculate the fraction of a turn the minute hand of a clock goes through from:

       i 7:15 to 7:50   **ii** 8:25 to 9:10   **iii** 6:24 to 7:12.

Problem solving: Shaded fraction

What fraction of this shape is shaded?

Give your fraction in its simplest form.

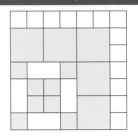

8.2 Comparing fractions

Learning objective

- To compare and order two or more fractions

Key word

fraction wall

You can compare fractions to each other, in terms of size. For example, $\frac{6}{8}$ is equivalent to $\frac{3}{4}$ but $\frac{2}{5}$ is bigger than $\frac{1}{4}$.

You can use a **fraction wall** to compare and order fractions.

$\frac{1}{8}$	$\frac{1}{8}$	$\frac{1}{8}$	$\frac{1}{8}$	$\frac{1}{8}$	$\frac{1}{8}$	$\frac{1}{8}$	$\frac{1}{8}$

| $\frac{1}{7}$ | $\frac{1}{7}$ | $\frac{1}{7}$ | $\frac{1}{7}$ | $\frac{1}{7}$ | $\frac{1}{7}$ | $\frac{1}{7}$ |

| $\frac{1}{6}$ | $\frac{1}{6}$ | $\frac{1}{6}$ | $\frac{1}{6}$ | $\frac{1}{6}$ | $\frac{1}{6}$ |

| $\frac{1}{5}$ | $\frac{1}{5}$ | $\frac{1}{5}$ | $\frac{1}{5}$ | $\frac{1}{5}$ |

| $\frac{1}{4}$ | $\frac{1}{4}$ | $\frac{1}{4}$ | $\frac{1}{4}$ |

| $\frac{1}{3}$ | $\frac{1}{3}$ | $\frac{1}{3}$ |

| $\frac{1}{2}$ | $\frac{1}{2}$ |

| 1 |

Example 3

Which is bigger, $\frac{3}{8}$ or $\frac{2}{5}$?

Compare the eighths and the fifths.

Counting from the left, the length occupied by two-fifths is slightly greater than the length occupied by three-eighths.

This means that $\frac{2}{5}$ is larger than $\frac{3}{8}$.

Example 4

Compare the two fractions $\frac{5}{6}$ and $\frac{7}{9}$.

Which one is larger? Give a reason for your answer.

The fraction wall does not have ninths so you cannot use that.

Instead, you need to work out equivalent fractions with the same denominator.

First, find a number that is a multiple of 6 and of 9.

The multiples of 6 are: 6, 12, 18, ...

The multiples of 9 are: 9, 18, 27, ...

18 is in both lists so it is a common multiple.

$\frac{5}{6} = \frac{15}{18}$ Multiply the numerator and denominator by 3.

$\frac{7}{9} = \frac{14}{18}$ Multiply the numerator and denominator by 2.

This shows that $\frac{5}{6}$ is larger than $\frac{7}{9}$.

Exercise 8B

1   Refer to the fraction wall.

Write down fractions that are equivalent to:

a $\frac{2}{3}$ **b** $\frac{6}{8}$

2   Refer to the fraction wall.

Write the correct sign (<, > or =) between the fractions in each pair.

a $\frac{4}{6}\cdots\frac{2}{3}$ **b** $\frac{5}{8}\cdots\frac{2}{3}$ **c** $\frac{3}{4}\cdots\frac{5}{7}$ **d** $\frac{2}{5}\cdots\frac{3}{7}$

e $\frac{5}{6}\cdots\frac{7}{8}$ **f** $\frac{1}{2}\cdots\frac{3}{8}$ **g** $\frac{3}{4}\cdots\frac{6}{7}$ **h** $\frac{7}{8}\cdots\frac{6}{7}$

3   Refer to the fraction wall to find a fraction that lies between the fractions in each pair.

a $\frac{1}{2}$ and $\frac{2}{3}$ **b** $\frac{2}{7}$ and $\frac{2}{5}$ **c** $\frac{5}{8}$ and $\frac{5}{7}$ **d** $\frac{2}{3}$ and $\frac{3}{4}$

4 Rewrite the fractions in each pair as fractions with the same denominator.

Then circle the smaller fraction.

a $\frac{1}{3}$ and $\frac{4}{9}$ **b** $\frac{2}{3}$ and $\frac{5}{9}$ **c** $\frac{1}{4}$ and $\frac{3}{10}$ **d** $\frac{3}{4}$ and $\frac{7}{10}$

5 Copy and complete each statement, using a fraction in its simplest form.

Refer to the fraction wall if you need to, to work out the missing fractions.

a ☐ is twice $\frac{1}{6}$ **b** ☐ is half of $\frac{1}{4}$ **c** ☐ is three times $\frac{1}{6}$ **d** ☐ is half of $\frac{3}{4}$

6 Write the fractions in each pair with the same denominator, to show which is larger.

a $\frac{2}{3}$ and $\frac{3}{4}$ **b** $\frac{2}{5}$ and $\frac{3}{10}$ **c** $\frac{2}{5}$ and $\frac{1}{4}$

d $\frac{2}{5}$ and $\frac{7}{20}$ **e** $\frac{3}{5}$ and $\frac{13}{20}$ **f** $\frac{1}{3}$ and $\frac{3}{10}$

7 Put the fractions in each set in order of size, starting with the smallest.

a $\frac{1}{2}, \frac{3}{4}, \frac{3}{5}$ **b** $\frac{2}{3}, \frac{5}{6}, \frac{3}{4}$ **c** $\frac{5}{8}, \frac{2}{3}, \frac{1}{2}$

8 Which of these fractions is nearer to 1: $\frac{5}{8}$ or $\frac{7}{10}$? Show all your working.

Investigation: The fractions in between

Suppose you are given two fractions.

You can find a fraction that lies between them – that is, larger than one of them but smaller than the other – by adding the numerators and adding the denominators to make a new fraction.

For example, suppose the fractions are $\frac{1}{2}$ and $\frac{5}{8}$.

Add the numerators: $1 + 5 = 6$

Add the denominators: $2 + 8 = 10$

This gives the new fraction $\frac{6}{10}$ which simplifies to $\frac{3}{5}$.

A Show that $\frac{3}{5}$ is larger than $\frac{1}{2}$ but smaller than $\frac{5}{8}$.

B Use this method to find a fraction between those in each pair.

Write your answer as simply as possible.

In each case, show that your new fraction is larger than one and smaller than the other.

a $\frac{1}{3}$ and $\frac{1}{6}$ **b** $\frac{2}{3}$ and $\frac{4}{5}$ **c** $\frac{2}{5}$ and $\frac{3}{4}$

8.3 Adding and subtracting fractions

Learning objective

- To add and subtract fractions with different denominators

Adding or subtracting fractions with the same denominator is easy. You just add or subtract the numerators and leave the denominator the same. You may be able to simplify the answer.

For example:

$$\frac{5}{8} + \frac{1}{8} = \frac{6}{8} = \frac{3}{4}$$

$$\frac{5}{8} - \frac{1}{8} = \frac{4}{8} = \frac{1}{2}$$

If the denominators are different, find an equivalent fraction to one or both, so that you have two fractions with the same denominator to add or subtract.

Example 5

Work these out.

a $\frac{2}{3} + \frac{1}{3}$ **b** $\frac{2}{3} - \frac{1}{6}$ **c** $\frac{2}{3} + \frac{1}{4}$

a $\frac{2}{3} + \frac{1}{3}$ The denominators are the same. Add the numerators.

$\frac{2}{3} + \frac{1}{3} = \frac{3}{3}$ Three thirds is one whole.

$\frac{2}{3} + \frac{1}{3} = \frac{3}{3} = 1$

b $\frac{2}{3} - \frac{1}{6}$

A common multiple of 3 and 6 is 6 so change $\frac{2}{3}$ to sixths by multiplying the denominator and numerator by 2.

$\frac{2}{3} - \frac{1}{6} = \frac{4}{6} - \frac{1}{6}$

$\qquad = \frac{3}{6} = \frac{1}{2}$

c $\frac{2}{3} + \frac{1}{4}$

A common multiple of 3 and 4 is 12.

Change both fractions to twelfths by multiplying the denominator and numerator of $\frac{2}{3}$ by 4 and the denominator and numerator of $\frac{1}{4}$ by 3.

$\frac{2}{3} + \frac{1}{4} = \frac{8}{12} + \frac{3}{12}$ $\frac{2}{3}$ is $\frac{8}{12}$ and $\frac{1}{4}$ is $\frac{3}{12}$.

$\qquad = \frac{11}{12}$ This cannot be simplified.

Exercise 8C

1 Add or subtract these fractions. Simplify the answers as much as possible.

a $\frac{3}{8} + \frac{1}{8}$ b $\frac{3}{10} + \frac{3}{10}$ c $\frac{5}{8} - \frac{3}{8}$ d $\frac{5}{12} - \frac{1}{12}$

e $\frac{7}{10} + \frac{3}{10}$ f $\frac{9}{10} - \frac{3}{10}$ g $\frac{5}{16} + \frac{3}{16}$ h $\frac{5}{16} - \frac{3}{16}$

2 Add these fractions. Write your answers as simply as possible.

a $\frac{3}{8} + \frac{1}{2}$ b $\frac{3}{8} + \frac{1}{4}$ c $\frac{5}{8} + \frac{1}{4}$ d $\frac{3}{4} + \frac{1}{8}$

e $\frac{1}{2} + \frac{3}{10}$ f $\frac{1}{3} + \frac{5}{12}$ g $\frac{1}{12} + \frac{3}{4}$ h $\frac{5}{16} + \frac{3}{8}$

3 Subtract these fractions. Write your answers as simply as possible.

a $\frac{1}{2} - \frac{1}{8}$ b $\frac{3}{8} - \frac{1}{4}$ c $\frac{5}{8} - \frac{1}{4}$ d $\frac{3}{4} - \frac{1}{8}$

e $\frac{1}{2} - \frac{3}{10}$ f $\frac{5}{12} - \frac{1}{3}$ g $\frac{3}{4} - \frac{1}{12}$ h $\frac{3}{8} - \frac{5}{16}$

4 Add or subtract these fractions. Write your answers as simply as possible.

a $\frac{1}{2} + \frac{1}{3}$ b $\frac{1}{2} - \frac{1}{3}$ c $\frac{3}{4} - \frac{1}{6}$ d $\frac{2}{3} + \frac{1}{4}$

e $\frac{2}{5} + \frac{1}{2}$ f $\frac{5}{8} - \frac{1}{3}$ g $\frac{3}{8} + \frac{1}{12}$ h $\frac{5}{6} - \frac{2}{9}$

5 Add these fractions.

a $\frac{3}{8} + \frac{5}{12}$ b $\frac{2}{5} + \frac{4}{9}$ c $\frac{3}{7} + \frac{1}{8}$ d $\frac{1}{2} + \frac{2}{5}$

e $\frac{8}{9} + \frac{1}{15}$ f $\frac{1}{6} + \frac{3}{8}$ g $\frac{7}{8} + \frac{1}{12}$ h $\frac{3}{10} + \frac{5}{12}$

6 Subtract these fractions.

a $\frac{7}{8} - \frac{5}{12}$ b $\frac{4}{5} - \frac{2}{9}$ c $\frac{3}{7} - \frac{1}{8}$ d $\frac{1}{2} - \frac{2}{5}$

e $\frac{7}{9} - \frac{4}{15}$ f $\frac{5}{6} - \frac{3}{8}$ g $\frac{7}{8} - \frac{1}{12}$ h $\frac{7}{10} - \frac{5}{12}$

7 In a book, $\frac{1}{12}$ of the space on the pages is taken up by photographs, $\frac{2}{3}$ is filled with text and the rest has diagrams.

What fraction of the book is taken up with diagrams? Give your answer in its simplest form.

8 A survey of the pupils in one class showed that $\frac{3}{5}$ of them walked to school, $\frac{1}{4}$ came by bus and rest came by car.

What fraction of the class came by car?

PS **9** John has £600. He gives $\frac{1}{4}$ to charity. He spends $\frac{2}{3}$ of it. He saves half of the rest. What fraction does he save?

PS **10** In a fridge there are two cartons of milk. One contains $\frac{2}{3}$ of a litre. The other contains $\frac{1}{4}$ of a litre.

a How much milk is there altogether?

b Milk is poured from one carton to the other so that they both contain the same amount.

How much milk was poured into the other carton? Give your answer as a fraction of a litre.

Challenge: Ancient Egyptians

The ancient Egyptians only used unit fractions.

These are fractions with a numerator of 1.

They would write $\frac{5}{8}$ as $\frac{1}{2} + \frac{1}{8}$.

A Write each of these fractions as a sum of two different unit fractions.

a $\frac{3}{8}$ **b** $\frac{5}{6}$ **c** $\frac{3}{10}$ **d** $\frac{4}{9}$

B Write this fraction as the sum of two unit fractions in two different ways.

$\frac{7}{12}$

C Write each of these fractions as the sum of **three** different unit fractions:

a $\frac{4}{5}$ **b** $\frac{9}{10}$

8.4 Mixed numbers and improper fractions

Learning objectives

- To convert mixed numbers to improper fractions
- To convert improper fractions to mixed numbers

Key words

convert

improper fraction

mixed number

This diagram shows the **mixed number** $2\frac{3}{4}$.

Each whole shape is divided into quarters. Altogether there are 11 quarters.

The whole could be written as $2\frac{3}{4}$ or $\frac{11}{4}$.

Fractions such as $\frac{11}{4}$ are called **improper fractions**. In an improper fraction the numerator is bigger than the denominator.

You can **convert** (change) improper fractions to mixed numbers and mixed numbers to improper fractions.

Example 6

a Convert $3\frac{2}{5}$ into an improper fraction.   **b** Convert $\frac{22}{3}$ into a mixed number.

a $3 = \frac{15}{5}$ 3×5 gives 15 fifths.

$3\frac{2}{5} = \frac{15}{5} + \frac{2}{5}$

$= \frac{17}{5}$

b $22 \div 3 = 7$ remainder 1 $\frac{22}{3}$ will give 7 whole ones plus a remainder of $\frac{1}{3}$.

$\frac{22}{3} = 7\frac{1}{3}$ Write the remainder as a fraction.

Addition of fractions can give an answer that is a mixed number.

Example 7

Work out $\frac{3}{4} + \frac{2}{3}$.

$\frac{3}{4} + \frac{2}{3} = \frac{9}{12} + \frac{8}{12}$ 4 and 3 are factors of 12 so change the fractions to twelfths

$= \frac{17}{12}$ This is an improper fraction. It cannot be simplified.

$= 1\frac{5}{12}$ $7 \div 12 = 1$ remainder 5.

Exercise 8D

1 Convert each of these mixed numbers to an improper fraction.

    a $2\frac{1}{2}$ **b** $4\frac{3}{4}$ **c** $3\frac{5}{6}$ **d** $7\frac{1}{4}$ **e** $5\frac{9}{10}$ **f** $6\frac{3}{8}$

    g $1\frac{11}{12}$ **h** $3\frac{7}{9}$ **i** $8\frac{3}{5}$ **j** $4\frac{4}{9}$ **k** $2\frac{4}{11}$ **l** $5\frac{6}{7}$

(MR) 2 Match the improper fractions to the mixed numbers.

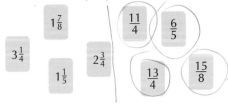

3 Convert each of these improper fractions into a mixed number.

    a $\frac{11}{5}$ **b** $\frac{25}{4}$ **c** $\frac{41}{5}$ **d** $\frac{29}{3}$ **e** $\frac{87}{10}$ **f** $\frac{51}{8}$

    g $\frac{43}{12}$ **h** $\frac{67}{9}$ **i** $\frac{35}{6}$ **j** $\frac{85}{7}$ **k** $\frac{18}{11}$ **l** $\frac{103}{12}$

(PS) 4 Sort these into order from smallest to largest.

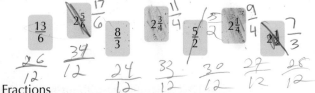

5 Which of these fractions is nearer in value to 1, $\frac{7}{8}$ or $\frac{8}{7}$?

Show how you decide.

(PS) **6** Write down a fraction with a value between $\frac{13}{6}$ and $\frac{13}{5}$.

(MR) **7** Look at each sequence, below.

a $\dfrac{17}{10}, \dfrac{17}{11}, \dfrac{17}{12}, \dfrac{17}{13}, \dfrac{17}{14}$ b $\dfrac{17}{10}, \dfrac{17}{9}, \dfrac{17}{8}, \dfrac{17}{7}, \dfrac{17}{6}$

i   Work out if the fractions in the sequence are increasing or decreasing in value.

ii   Give a reason for your answer.

(PS) **8** Look at each sequence, below.

a $\dfrac{17}{10}, \dfrac{18}{11}, \dfrac{19}{12}, \dfrac{20}{13}, \dfrac{21}{14}$ b $\dfrac{17}{10}, \dfrac{16}{9}, \dfrac{15}{8}, \dfrac{14}{7}, \dfrac{13}{6}$

i   Work out if the fractions in the sequence are increasing or decreasing in value.

ii   Give a reason for your answer.

Challenge: Fill the bags

At a builders' merchant, sand is normally packed in 5 kg bags.

Three bags have fallen over and spilled their contents.

One bag is $\frac{3}{4}$ full. A second is $\frac{2}{3}$ full. A third is $\frac{5}{8}$ full.

Is there enough sand to fill two bags completely?
Give a reason for your answer.

8.5 Calculations with mixed numbers

Learning objective

• To add and subtract simple mixed numbers with different denominators

When adding and subtracting mixed numbers, you can work with the whole-number parts and then with the fraction parts, before combining them to form your answer.

Example 8

Work these out.

a $4\frac{2}{3} + 2\frac{1}{4}$ b $5\frac{3}{4} + 2\frac{5}{6}$

$\quad$ a $\;4\frac{2}{3} + 2\frac{1}{4} = 4 + \frac{2}{3} + 2 + \frac{1}{4}$ Think of $4\frac{2}{3}$ as $4 + \frac{2}{3}$ and $2\frac{1}{4}$ as $2 + \frac{1}{4}$.

$\qquad\qquad\qquad = 4 + 2 + \frac{2}{3} + \frac{1}{4}$ Add the whole numbers and the fractions separately.

$\qquad\qquad\qquad = 6\frac{11}{12}$ $\frac{2}{3} + \frac{1}{4} = \frac{8}{12} + \frac{3}{12} = \frac{11}{12}$

(Continued)

b $5\frac{3}{4} + 2\frac{5}{6} = 5 + \frac{3}{4} + 2 + \frac{5}{6}$

$= 7 + \frac{3}{4} + \frac{5}{6}$

Now $\frac{3}{4} + \frac{5}{6} = \frac{9}{12} + \frac{10}{12} = \frac{19}{12}$ 12 is a common multiple of 4 and 6.

$= 1\frac{7}{12}$ This is what to do if the fractions add to more than 1.

So $5\frac{3}{4} + 2\frac{5}{6} = 7 + 1\frac{7}{12}$

$= 8\frac{7}{12}$

You can subtract mixed numbers in a similar way.

Example 9 shows how to subtract the fractions in Example 8.

Example 9

Work these out.

a $4\frac{2}{3} - 2\frac{1}{4}$ **b** $5\frac{3}{4} - 2\frac{5}{6}$

a $4\frac{2}{3} - 2\frac{1}{4} = 4 + \frac{2}{3} - 2 - \frac{1}{4}$ Subtract 2 and subtract $\frac{1}{4}$.

$= 4 - 2 + \frac{2}{3} - \frac{1}{4}$ Subtract the whole numbers and the fractions separately.

$= 4 - 2 + \frac{8}{12} - \frac{3}{12}$

$= 2\frac{5}{12}$

b $5\frac{3}{4} - 2\frac{5}{6} = 5 + \frac{3}{4} - 2 - \frac{5}{6}$

$= 5 - 2 + \frac{3}{4} - \frac{5}{6}$

$= 3 + \frac{3}{4} - \frac{5}{6}$

Now $\frac{3}{4} - \frac{5}{6} = \frac{9}{12} - \frac{10}{12} = -\frac{1}{12}$ A negative answer because $\frac{3}{4}$ is less than $\frac{5}{6}$.

This means that $5\frac{3}{4} - 2\frac{5}{6} = 3 - \frac{1}{12}$ Subtract $\frac{1}{12}$ from 3.

$= 2\frac{11}{12}$

Another method, that avoids negative fractions, is to write both numbers as improper fractions.

$5\frac{3}{4} - 2\frac{5}{6} = \frac{23}{4} - \frac{17}{6}$ $5 \times 4 + 3 = 23$ and $2 \times 6 + 5 = 17$

$= \frac{69}{12} - \frac{34}{12}$ Change the quarters and sixths to twelfths.

$= \frac{35}{12}$

$= 2\frac{11}{12}$ $35 \div 12 = 2$ remainder 11.

You can use either method.

Exercise 8E

Write your answers in their simplest form.

1   Add the mixed numbers.

     a $1\frac{2}{3} + 2\frac{2}{3}$ **b** $2\frac{1}{3} + \frac{5}{6}$ **c** $4\frac{3}{10} + 2\frac{1}{5}$ **d** $3\frac{7}{9} + 5\frac{4}{9}$

     e $3\frac{8}{15} + 2\frac{7}{15}$ **f** $2\frac{7}{9} + 3\frac{13}{18}$ **g** $4\frac{5}{12} + 1\frac{1}{3}$ **h** $3\frac{9}{14} + 3\frac{6}{7}$

2 Subtract the mixed numbers.

a $5\frac{9}{10} - 2\frac{7}{10}$ **b** $2\frac{1}{6} - \frac{5}{6}$ **c** $3\frac{5}{7} - 1\frac{6}{7}$ **d** $4\frac{8}{9} - 1\frac{5}{9}$

e $6\frac{4}{15} - 5\frac{8}{15}$ **f** $4\frac{4}{9} - 2\frac{7}{9}$ **g** $8\frac{5}{12} - 2\frac{11}{12}$ **h** $5\frac{7}{10} - 1\frac{9}{10}$

3 Add or subtract these mixed numbers.

a $4\frac{1}{2} + 1\frac{1}{3}$ **b** $4\frac{1}{2} - 1\frac{1}{3}$ **c** $5\frac{3}{4} + 3\frac{1}{3}$ **d** $5\frac{3}{4} - 3\frac{1}{3}$

e $4 + 2\frac{3}{5}$ **f** $4 - 2\frac{3}{5}$ **g** $9\frac{2}{3} + 5\frac{3}{5}$ **h** $9\frac{2}{3} - 5\frac{3}{5}$

4 Work out the sum or difference.

a $5\frac{5}{8} + 3\frac{1}{16}$ **b** $1\frac{3}{4} + 6\frac{7}{10}$ **c** $9\frac{1}{3} - \frac{11}{12}$ **d** $2\frac{1}{2} - 1\frac{2}{3}$

e $4\frac{7}{10} + 4\frac{4}{5}$ **f** $6\frac{2}{3} - 1\frac{1}{8}$ **g** $12\frac{3}{5} + 13\frac{1}{8}$ **h** $6\frac{2}{5} - 5\frac{3}{4}$

(PS) **5** Work out the missing number in each calculation.

a $2\frac{1}{3} + ? = 5$ **b** $2\frac{1}{4} + ? = 7\frac{7}{8}$ **c** $4\frac{2}{3} + ? = 6\frac{1}{5}$

6 Jason is carrying five bags. He has two, each, of mass $3\frac{1}{2}$ kg, in his left hand, and three, each of mass $2\frac{2}{5}$ kg, in his right hand.

a Which hand is carrying the heavier load? Show your working.

b What is the mass of the five bags altogether?

7 Laura lives $5\frac{1}{4}$ miles from school. Jim lives $2\frac{2}{3}$ miles from school.

a How much further from school does Jim live than Laura?

 b What is the maximum distance there could be between where Jim and Laura live?

8 At birth, triplets weighed $5\frac{3}{4}$ pounds, $5\frac{1}{2}$ pounds and $5\frac{1}{4}$ pounds. Four months later, they each weighed 11 pounds.

a How much did they weigh at birth altogether?

b How much has their total weight increased in four months?

Challenge: Fractions in sequences

These are the first four numbers in a sequence.

$1\frac{1}{2}$, $2\frac{2}{3}$, $3\frac{3}{4}$, $4\frac{4}{5}$, ...

A Write down the fifth and sixth numbers in the sequence.

B Work out the difference between:

a the first and second numbers

b the second and third numbers

c the third and fourth numbers.

C a Predict what the difference between the fifth and sixth numbers will be.

b Do a calculation to check whether your prediction is correct.

Ready to progress?

Review questions

1   Here are two fractions.

   $\dfrac{5}{6}$ $\dfrac{3}{8}$

   a   Write fractions equivalent to $\frac{5}{6}$ and $\frac{3}{8}$ that have the same numerator.

   b   Write fractions equivalent to $\frac{5}{6}$ and $\frac{3}{8}$ that have the same denominator.

   c   Add the two fractions.

   d   Work out the difference between the two fractions.

2   Write each fraction as a mixed number in its simplest form.

   a $\dfrac{20}{15}$ b $\dfrac{20}{16}$ c $\dfrac{20}{12}$ d $\dfrac{20}{8}$ e $\dfrac{20}{3}$

3   Here are four numbers.

   $2\frac{3}{4}$ $\frac{11}{5}$ $2\frac{3}{5}$ $\frac{9}{4}$

   Write the numbers in order, smallest first.

4   Work these out.

   a $\dfrac{2}{3}+\dfrac{1}{2}$ b $\dfrac{2}{3}-\dfrac{1}{2}$ c $\dfrac{3}{4}+\dfrac{2}{3}$ d $\dfrac{3}{4}-\dfrac{2}{3}$ e $\dfrac{4}{5}+\dfrac{3}{4}$ f $\dfrac{4}{5}-\dfrac{3}{4}$

5   This is the start of a sequence of numbers.

   $1\frac{2}{7}, \ 1\frac{1}{2}, \ 1\frac{5}{7}, \ 1\frac{13}{14}, \dots$

   a   The difference between adjacent numbers is always the same. What is the difference?

   b   Work out the next number in the sequence. Write the fraction as simply as possible.

 6   Work out the number that is halfway between the two numbers shown on each scale.

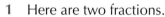

   $\frac{1}{2}$ $\frac{2}{3}$ $\frac{1}{6}$ $\frac{11}{18}$ $1\frac{2}{3}$ $2\frac{1}{5}$

7   Work out the perimeter of this rectangle.

Give your answer as a mixed number.

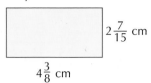 $2\frac{7}{15}$ cm

$4\frac{3}{8}$ cm

8   Work out the perimeter of this L-shape. Give your answer as a mixed number.

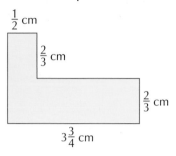

$\frac{1}{2}$ cm

$\frac{2}{3}$ cm

$\frac{2}{3}$ cm

$3\frac{3}{4}$ cm

9   Work out the mode, median and range for this set of data.

$14\frac{3}{5}$ kg $9\frac{2}{3}$ kg $15\frac{1}{5}$ kg $15\frac{3}{8}$ kg $14\frac{3}{5}$ kg $16\frac{3}{4}$ kg

10 Simplify this expression.

$4\frac{1}{4}x + 3\frac{3}{4}y + 2\frac{1}{2}x - 1\frac{1}{8}y$

11 This is the start of a sequence of numbers.

$3\frac{1}{2}$ $5\frac{5}{12}$ $7\frac{1}{3}$ $9\frac{1}{4}$

Work out:

a   the sum of the first and the fourth terms

b   the sum of the second and the third terms

c   the difference between the first and the second terms

d   the difference between the third and the fourth terms

e   the next two numbers in the sequence.

12 This sum is correct.

$3\frac{2}{5} + 4\frac{3}{4} + \boxed{} = 10$

Work out the missing number.

Challenge

Fractional dissections

Task 1

This rectangle has 12 squares.
It is divided into three parts.

a Check that the red part is $\frac{1}{4}$, the yellow part is $\frac{1}{6}$ and the green part is $\frac{7}{12}$.

This shows that $\frac{1}{4} + \frac{1}{6} + \frac{7}{12} = 1$.

This is one way to write the whole rectangle as the sum of three fractions with different denominators (in this case 4, 6 and 12). We can call this a triple dissection of the rectangle.

b Draw diagrams that show two different triple dissections. Each one should have a different set of three fractions. Do not simply make a different drawing of the same triple dissection.

Task 2

Draw diagrams to show all the different possible triple dissections for each of these rectangles.

In each case, there may be none, one or more than one.

a   b   c   d   e

Task 3

Find all the possible triple dissections of this rectangle.

Are you sure you have found them all?

Task 4

A unit fraction has a numerator of 1.

This rectangle has been divided into four parts.

You should see that $\frac{1}{2} + \frac{1}{4} + \frac{1}{6} + \frac{1}{12} = 1$.

They are all unit fractions and they are all different.

This rectangle can be divided into different unit fractions in several ways.

Find the way that uses the largest possible number of different unit fractions.

9

Angles

This chapter is going to show you:

- how to measure and draw angles
- how to calculate angles at a point, angles on a straight line and vertically opposite angles
- how to calculate angles in a triangle
- how to calculate angles in a quadrilateral
- how to recognise parallel, intersecting and perpendicular lines
- how to calculate angles in parallel lines
- how to explain the geometrical properties of triangles and quadrilaterals
- how to use algebra to work out the size of angles.

You should already know:

- the names of the different types of angle
- the names of different triangles and quadrilaterals
- how to solve an equation.

About this chapter

Why does light travel down optical fibres made of glass, like these?

The answer is found in angles. As light travels down the fibre it is constantly reflected back from the sides, instead of passing out of the fibre. The light has to hit the sides of the fibre at an angle that is greater than a certain angle, called the critical angle. Fibre optics are now used for ultrafast telephone and broadband communications throughout the world. So that is one very important angle!

9.1 Measuring and drawing angles

Learning objectives

- To use a protractor to measure an angle
- To use a protractor to draw an angle

Key words	
acute angle	angle
degrees	obtuse angle
protractor	reflex angle
right angle	

When two lines meet at a point they form an **angle**.
An angle is measured in **degrees**.

The types of angle are shown below.

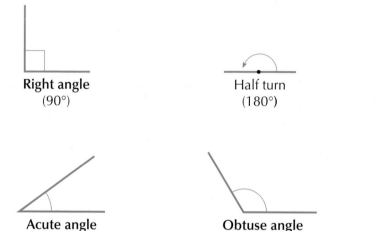

Right angle
(90°)

Half turn
(180°)

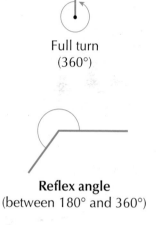

Full turn
(360°)

Acute angle
(less than 90°)

Obtuse angle
(between 90°
and 180°)

Reflex angle
(between 180° and 360°)

When you need to measure angles, you use a **protractor**.

Notice that there are two scales on a protractor.

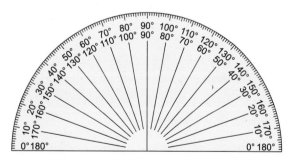

The outer scale goes from 0° to 180° and the inner one goes from 180° to 0°.

It is important that you use the correct scale.

When measuring or drawing an angle, always decide first whether it is an acute angle (less than 90°) or an obtuse angle (more than 90°).

Example 1

Measure this angle.

First, decide whether the angle you are measuring is acute or obtuse.

This is an acute angle (less than 90°).

Place the centre of the protractor exactly over the point of the angle, as in the diagram.

The two angles shown on the protractor scales are 60° and 120°. Since you are measuring an acute angle, the angle is 60° (to the nearest degree).

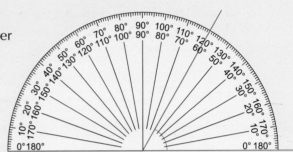

Example 2

Measure this angle.

First, decide whether the angle you are measuring is acute or obtuse.

This is an obtuse angle (greater than 90°).

Place the centre of the protractor exactly over the corner of the angle, as in the diagram.

The two angles shown on the protractor scales are 40° and 140°.

Since you are measuring an obtuse angle, the angle is 140° (to the nearest degree).

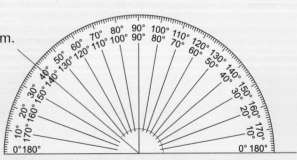

Example 3

Draw and label an angle of 70°.

First draw a line about 5 cm long.

Place the middle of the protractor exactly over one end of the line.

Put a mark on 70°, as it is an acute angle.

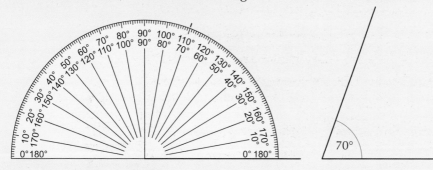

Then complete the angle.

Example 4

Measure this reflex angle.

First, measure the inside or interior angle. This is an obtuse angle.

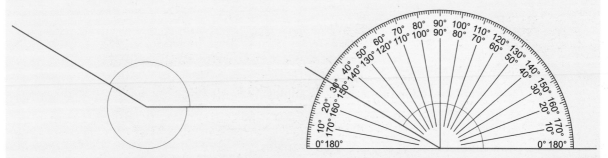

The two angles shown on the protractor scales are 30° and 150°.

Since you are measuring an obtuse angle, the angle is 150°.

You can find the size of the reflex angle by subtracting this angle from 360°.

The reflex angle is therefore 360° − 150°, which is 210° (to the nearest degree).

1 Measure each of these angles, giving your answer to the nearest degree.

a

b

c

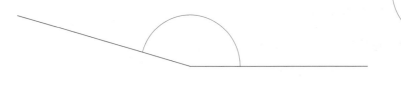

d

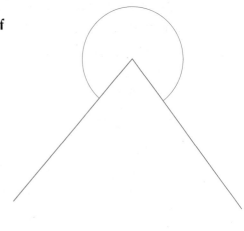

e

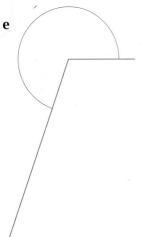

f

2 Draw and label each of these acute angles.

　a 30°　　　　　**b** 45°　　　　　**c** 63°　　　　　**d** 87°

3 Draw and label each of these obtuse angles.

　a 110°　　　　　**b** 137°　　　　　**c** 162°　　　　　**d** 175°

4 Draw and label each of these reflex angles.

　a 190°　　　　　**b** 225°　　　　　**c** 285°　　　　　**d** 300°

5 **a** Measure the three angles in this triangle.

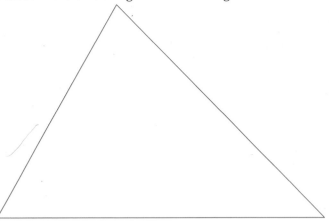

b Add the three angles together.

6 Draw a triangle with an obtuse angle, as in this sketch.

a Measure the three angles in your triangle.

b Add the three angles together.

c You should now be able to complete this statement:

In any triangle the sum of the three angles is always … .

7 **a** Measure the five angles in this pentagon.

b Add the five angles together.

c Can you see a connection between this answer and the sum of the angles of the triangle?

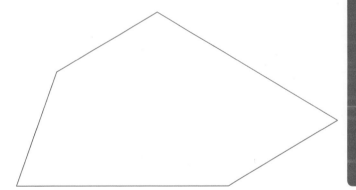

Challenge: Measuring angles

A Draw an acute angle.

Ask a partner to estimate the size of the angle.

Was the estimate close to the exact answer?

Repeat this challenge three times.

B Repeat this challenge for three obtuse angles.

C Repeat this challenge for three reflex angles.

9.2 Calculating angles

Learning objectives

- To understand the properties of parallel, intersecting and perpendicular lines
- To calculate angles at a point
- To calculate angles on a straight line
- To calculate opposite angles

You can **calculate** (work out) the unknown angles in a diagram from the information given.

Unknown angles are usually denoted by letters, such as $a, b, c, \ldots$.

Remember that diagrams are not usually drawn to scale.

Key words

angles at a point

angles on a straight line

calculate

intersect

opposite angles

parallel

perpendicular

vertically opposite angles

Describing lines

A straight line can be considered to have infinite length, that is, it can go on forever in both directions.

A segment of a straight line has a fixed (finite) length.

A ————————————————— B

The line AB has two end points, one at A and the other at B.

Any two lines are either **parallel** or they **intersect** (cross).

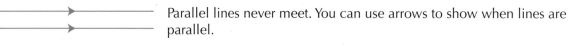

Parallel lines never meet. You can use arrows to show when lines are parallel.

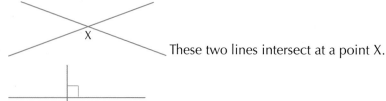

These two lines intersect at a point X.

These two lines intersect at right angles.
The lines are **perpendicular**.

Right angles

In the diagram, the square symbol means that the angle is 90° or a **right angle**.

Example 5

Calculate the size of the angle marked x.

$x = 90° - 48°$

$x = 32°$

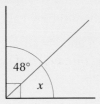

Angles at a point add up to 360°.

In this diagram $a + b + c = 360°$.

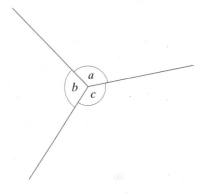

Example 6

Calculate the size of the angle marked a.

The three angles add up to 360° so:

$a = 360° - 150° - 130°$

$a = 80°$

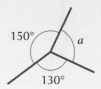

Angles on a straight line

Angles on a straight line add up to 180°.

In this diagram $a + b = 180°$.

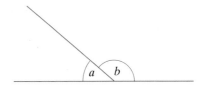

Example 7

Calculate the size of the angle marked b.

The two angles on a straight line add up to 180° so:

$b = 180° - 155°$

$b = 25°$

Vertically opposite angles

When two lines intersect, the **vertically opposite angles** are equal.

You will often see them simply called **opposite angles**.

In the diagram, $a = d$ and $b = c$.

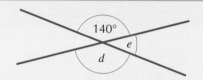

Example 8

Calculate the sizes of the angles marked d and e.

Give reasons for your answers.

$d = 140°$ Opposite angles are equal.

Angles on a straight line = 180° so:

$e = 180° - 140°$

  $= 40°$

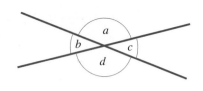

Exercise 9B

1   Calculate the size of each unknown angle.

a   **b**   **c**   **d**

 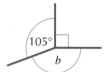 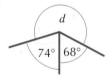

2   Calculate the size of each unknown angle.

a   **b**   **c**   **d**

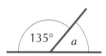 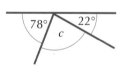 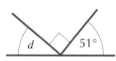

3   Calculate the size of each unknown angle.

a   **b**   **c**   **d**

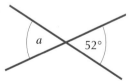 

4 Calculate the size of each unknown angle.

Explain how you worked out your answers.

a

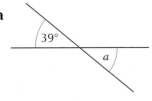

39°
a

b

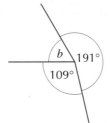

b 191°
109°

c

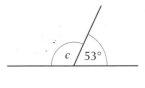

c 53°

d

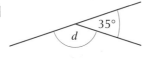

35°
d

e

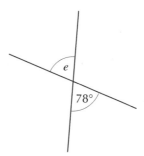

e
78°

f

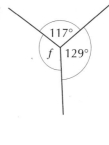

117°
f 129°

(MR) 5 Calculate the size of each unknown angle.

Explain how you worked out your answers.

a

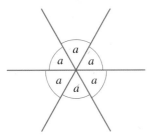

a
a *a*
a *a*
á

b

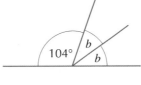

104° *b*
b

c

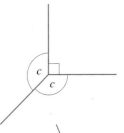

c
c

6 In the diagram the angle marked *d* is twice the size of the angle marked *e*.

Calculate the size of each angle.

7 In the diagram, *x* and *y* lie on a straight line and the value of *y* is 20° more than the value of *x*.

Calculate a pair of angles to make this statement true.

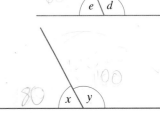

120
60
e *d*

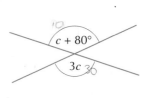

100
80 *x* *y*

Reasoning: Calculating angles

Calculate the size of each of the angles marked with letters.

Explain your answers.

A

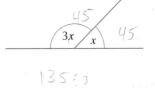

45
3*x* *x* 45
135÷3
120

B

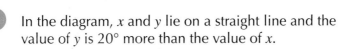

20°
y
220° 55
4*y* 2*y* 5
3*y* 30
90°

C

10
30°
c + 80°
3*c* 30

9.3 Corresponding and alternate angles

Learning objective

* To calculate angles in parallel lines

Key words

alternate angles

corresponding angles

transversal

A line that intersects a set of parallel lines is called a **transversal**.

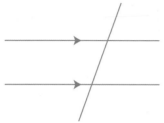

Notice in the diagram that eight distinct angles are formed by a transversal.

The two angles marked on this diagram are equal and are called **corresponding angles**.

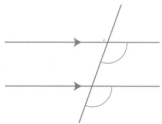

You may find it helpful to remember which angles are corresponding angles by noting that they occur where the lines make the shape of the letter F.

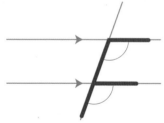

The two angles marked on this diagram are equal and are called **alternate angles**.

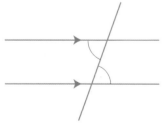

Look for where the lines make the shape of the letter Z to identify alternate angles.

Example 9

Calculate the sizes of the angles marked p and q.

Give reasons for your answers.

a

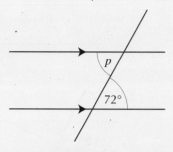

b

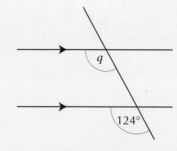

a $p = 72°$ (Alternate angles are equal.) **b** $q = 124°$ (Corresponding angles are equal.)

Exercise 9C

1 Which diagrams show a pair of alternate angles?

a

b

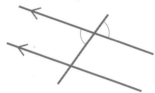

c

d

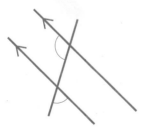

e

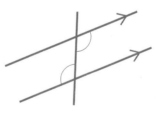

f

2 Which diagrams show a pair of corresponding angles?

a

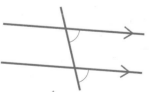

b

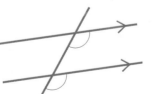

c

d

e

f

3 Calculate the size of each unknown angle.

State whether it is an alternate angle or a corresponding angle.

a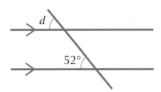

b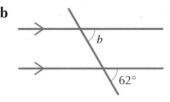

c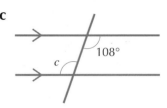

d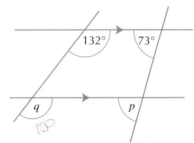

e

f

4 Calculate the size of each unknown angle.

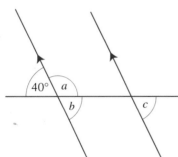

(MR) 5 Calculate the size of each unknown angle.

Explain how you worked out your answers.

a

b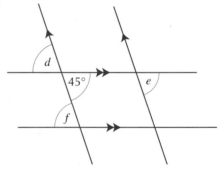

(MR) 6 **a** Calculate the size of each unknown angle.

Explain how you worked out your answers.

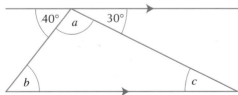

b What do you notice about the three angles in the triangle?

7 Draw a diagram similar to this one.

Measure the acute angle shown at the point A.

Draw the corresponding angle at the point marked B.

Extend the lines of your two angles so that you have a pair of parallel lines.

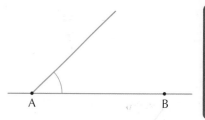

Reasoning: Angles in parallel lines

A Calculate the size of each unknown angle.

Explain how you worked out your answers.

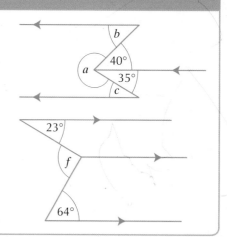

B Calculate the size of the angle marked *f* on the diagram.

Draw a diagram to explain your answer.

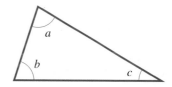

9.4 Angles in a triangle

Learning objective

• To know that the sum of the angles in a triangle is 180°

The angles in a **triangle** add up to 180°.

$a + b + c = 180°$

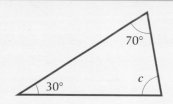

Example 10

Calculate the size of the angle marked *c*.

$c = 180° − 70° − 30°$

$c = 80°$

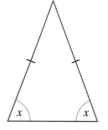

This is an **isosceles** triangle.

The two sides marked with dashes are the same length.

The two angles marked *x* are the same size.

Example 11

Calculate the size of the unknown angles in this isosceles triangle.

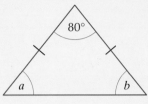

$180° - 80° = 100°$

Since a and b are the same, $100° ÷ 2 = 50°$.

So $a = b = 50°$.

You can also use the angles inside a triangle to calculate an angle outside it.

Example 12

Calculate the sizes of the angles marked a and b.

Give reasons for your answers.

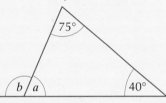

$a = 180° - 75° - 40°$

So $a = 65°$.   (Sum of angles in a triangle = 180°)

$b = 180° - 65°$

So $b = 115°$.   (Sum of angles on a line = 180°)

In the diagram in Example 12, the angle marked a is called the **interior angle** of the triangle because it is inside the triangle. The angle marked b is called the **exterior angle** of the triangle because it is outside the triangle.

Exercise 9D

1 Calculate the size of the unknown angle in each diagram.

a

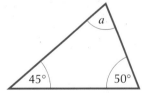

b

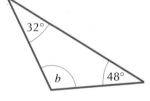

c

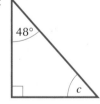

d

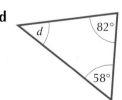

e

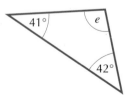

2 Calculate the size of the unknown angle in each diagram.

a
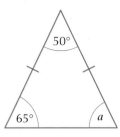

b
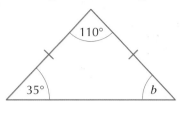

c
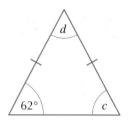

d
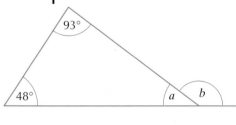

e
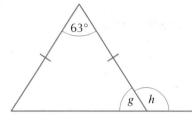

3 **a** Calculate the size of the unknown angles in each diagram.

b State whether the unknown angles are interior or exterior angles.

i
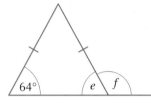

ii
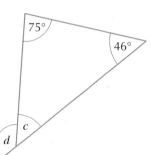

iii
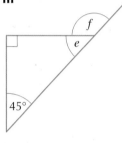

 c What do you notice about the exterior angles and the two known interior angles?

(PS) **4** In this triangle, the value of b is 5° more than the value of a.
The value of c is 5° more than the value of b.

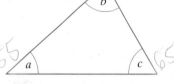

What are the sizes of the three angles?

(MR) **5** One angle in an isosceles triangle is 25°.
Calculate the possible sizes of the other two angles.

Explain, with diagrams, how you worked out your answer.

6 Calculate the size of each unknown angle.

a

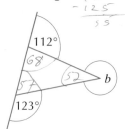

68°

360
−220
140

a

b

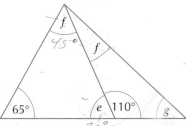

112°
68
52 *b*
57
123°

180
−125
55

c

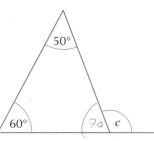

180
−110
70

50°

60° 70° *c*

d

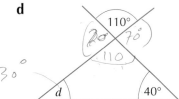

110°
20 70
110

30°

d 40°

e

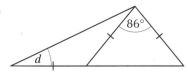

f
45°
f

65° *e* 110° *g*
70°

 7 Copy each diagram, then label with a letter each of the angles you need to know, to calculate the unknown angles *c* and *d*.

Give a reason for each angle.

a

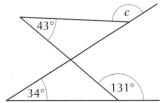

43° *c*

34° 131°

b

86°

d

 8 Calculate the sizes of the unknown angles in each diagram.

a

z=30
2z
z·6
4z z=15 3z 20°

b

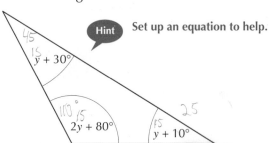

45
15
y + 30°

Hint **Set up an equation to help.**

110°
15
2*y* + 80°
15
y + 10°
25

Problem solving: Calculating angles in triangles

Calculate the size of the unknown angle in each diagram.

Copy each diagram and label it to explain, with reasons, how you obtained your answers.

A

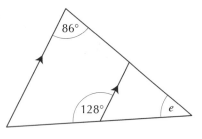

86°

128° *e*

B

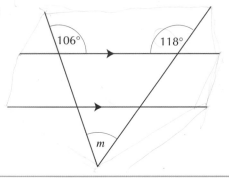

106° 118°

m

9.5 Angles in a quadrilateral

Learning objective

* To know that the sum of the angles in a quadrilateral is 360°

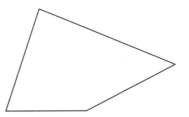

Key word

quadrilateral

A **quadrilateral** is any shape with four straight sides.

Draw a large quadrilateral similar to this one.

Use a protractor to measure each of the four angles.

Now add up the four angles. What do you notice?

You should find that your answer is close to 360°.

Now draw a different quadrilateral and find the sum of the four angles.

How close were you to 360°?

The angles in a quadrilateral add up to 360°.

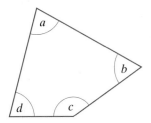

So, in the diagram: $a + b + c + d = 360°$.

Example 13

Calculate the sizes of the angles marked a and b on the diagram.

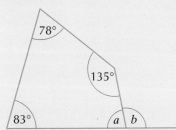

The angles in a quadrilateral add up to 360°.

So $a = 360° - 135° - 78° - 83°$

$\quad = 64°$

The angles on a straight line add up to 180°.

So $b = 180° - 64°$

$\quad = 116°$

1 In each quadrilateral, calculate the size of each angle marked by a letter.

a

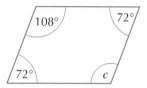

b

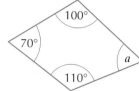

c

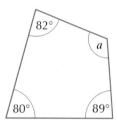

d

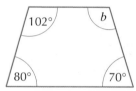

2 In each quadrilateral, calculate the size of each angle marked by a letter.

a

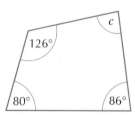

b

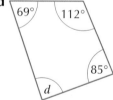

c

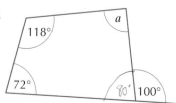

d

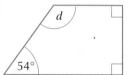

3 Which of these three sets of angles could be the four interior angles of a quadrilateral?

Explain how you worked out your answer.

a 60°, 70°, 80° and 140°   **b** 61°, 95°, 95° and 109°   **c** 41°, 89°, 112° and 128°

4 In each diagram, calculate the size of the unknown angle.

Explain how you worked out your answers.

a

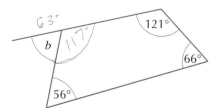

b

5   The diagram shows a trapezium.

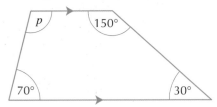

a   Calculate the size of the angle marked p.

(MR)   **b**   What do you notice about the angles in the trapezium?

(PS) 6   Calculate the size of the four angles in the quadrilateral.

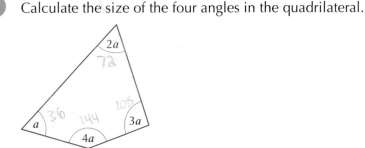

Explain how you worked out your answer.

(PS) 7   The four angles in a quadrilateral are a, $a + 10°$, $a + 20°$ and $a + 30°$.

Set up an equation and solve it to calculate the four angles.

(MR) 8   Can a quadrilateral have exactly three right angles?

Give a reason for your answer.

Problem solving: Triangles in quadrilaterals

A quadrilateral can be split into two triangles, as shown in the diagram.

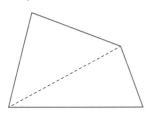

Explain how you can use this fact to show that the sum of the angles in any quadrilateral is 360°.

9.6 Properties of triangles and quadrilaterals

Learning objectives

- To understand and use the properties of triangles
- To understand and use the properties of quadrilaterals

Describing angles

The angle at B can be written as:

∠B or ∠ABC or angle ABC.

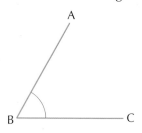

Describing triangles and quadrilaterals

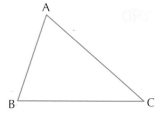

This triangle can be described as triangle ABC.

Each corner is called a **vertex**.

So it has:

- three **vertices**, A, B and C
- three angles, ∠A, ∠B and ∠C
- three sides, AB, AC and BC.

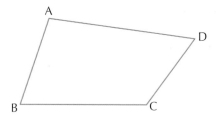

The quadrilateral can be described as quadrilateral ABCD.

So it has:

- four vertices, A, B, C and D
- four angles, ∠A, ∠B, ∠C and ∠D
- four sides, AB, BC, CD and AD.

In some shapes some or all of the sides are the same size or may be parallel to each other, or some or all angles the same. These are called their **geometrical properties**.

Types of triangle

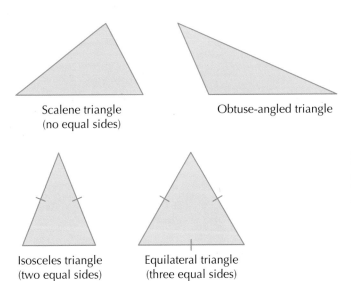

Scalene triangle
(no equal sides)

Obtuse-angled triangle

Right-angled triangle

Isosceles triangle
(two equal sides)

Equilateral triangle
(three equal sides)

Example 14

Describe the geometrical properties of the isosceles triangle ABC.

$$AB = AC$$
$$\angle ABC = \angle ACB$$

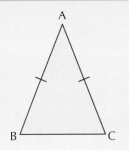

Types of quadrilateral

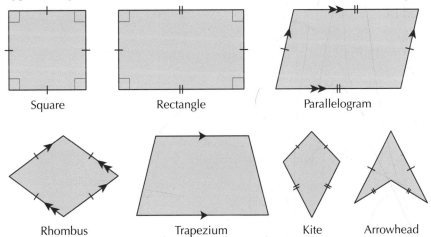

Square

Rectangle

Parallelogram

Rhombus

Trapezium

Kite

Arrowhead

Example 15

Describe the geometrical properties of the parallelogram ABCD.

AB = CD and AD = BC.

AB is parallel to CD.

AD is parallel to BC.

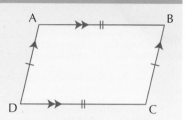

Exercise 9F

1 Describe the geometrical properties of the equilateral triangle ABC.

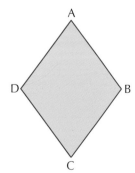

2 Look at the quadrilateral ABCD.

  a Write down two lines that are equal in length.

  b Write down two lines that are parallel.

  c Write down two lines that are perpendicular to each other.

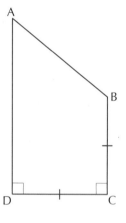

3 Describe the geometrical properties of the rhombus ABCD.

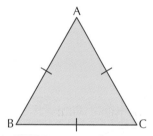

4 Which quadrilaterals have the following properties?

    a Four equal sides

    b Two different pairs of equal sides

    c Two pairs of parallel sides

    d Only one pair of parallel sides

    e Adjacent sides equal

    f Diagonals equal in length

    g Diagonals bisecting each other

    h Diagonals perpendicular to each other

    i Diagonals intersecting at right angles outside the shape

MR **5** Explain the difference between:

    a a square and a rectangle

    b a rhombus and a parallelogram

    c a kite and an arrowhead.

6 Copy this square on a piece of card.

Then draw in the two diagonals and cut out the four triangles.

How many different triangles or quadrilaterals can you make with:

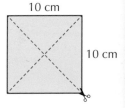

10 cm

10 cm

    a four of the triangles   **b** three of the triangles

    c two of the triangles?

7 With a partner, make a list of at least 10 examples of quadrilaterals you see in everyday life. If you can, take photos or make drawings. Describe them as 'square', 'rectangle', 'parallelogram', 'rhombus', 'trapezium', 'kite' or 'arrowhead'.

Use the internet for some ideas.

Suggest reasons why these quadrilaterals are used instead of other shapes.

Challenge: Making triangles and quadrilaterals

How many distinct triangles can you construct on this 3 by 3 pin-board?

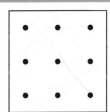

A Use square dotted paper to record your triangles.

Below each one, write down what type of triangle it is.

Here is an example.

This is a right-angled triangle.

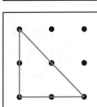

B How many distinct quadrilaterals can you make on the same 3 by 3 grid?

Ready to progress?

I can draw and measure angles.
I can recognise parallel, intersecting and perpendicular lines.
I can calculate angles at a point, angles on a straight line, vertically opposite angles and angles in a triangle.
I can describe and use the properties of different triangles.

I can calculate angles in parallel lines.
I can describe and use the properties of different quadrilaterals.

Review questions

1   Calculate the size of each unknown angle. Give reasons for your answers.

a

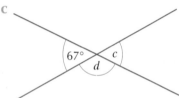

b

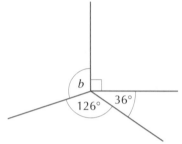

c

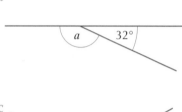

2   In each triangle, calculate the size of the unknown angles.

a

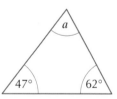

b

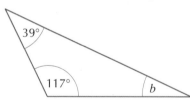

c

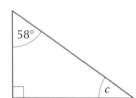

d

3   State whether the following statements are true or false.

a   A triangle can have three acute angles.

b   A triangle can have three obtuse angles.

c   A triangle can have two acute angles and one obtuse angle.

d   A triangle can have one acute angle and two obtuse angles.

4 The diagram shows triangle PQR.

Calculate the sizes of the angles marked *a*, *b* and *c*.

Give a reason for each answer.

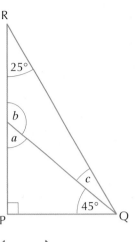

5 Quadrilateral A and quadrilateral B fit together to form a right-angled triangle.

Calculate the size of the angles marked *a*, *b*, *c* and *d* in quadrilateral B.

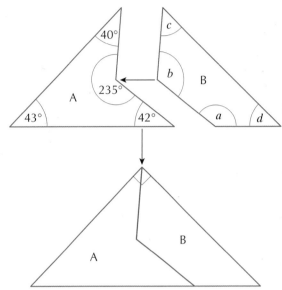

6 Calculate the sizes of the angles marked with letters on the diagram.

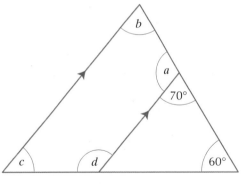

7 In the diagram, triangle DEC is drawn inside the parallelogram ABCD.

∠AED = 54° and ∠DCE = 48°.

Calculate the size of ∠DEC.

Give reasons for your answer.

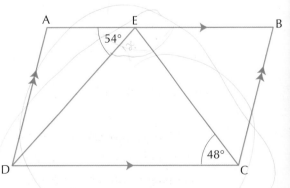

Activity

Constructing triangles and quadrilaterals

You need to be able to use a ruler, a protractor and compasses to draw shapes exactly, from the information you are given. This is called constructing a shape.

When constructing a shape you must draw lines accurately to the nearest millimetre and angles to the nearest degree.

Example: How to construct a triangle if you know two angles and one side.

Construct the triangle XYZ.

- Draw a line YZ 8.3 cm long.

- Draw an angle of 42° at Y.

- Draw an angle of 51° at Z.

- Extend both angle lines to intersect at X to complete the triangle.

The completed, full-sized triangle is given below.

Nowadays many architects and draughtsmen use computer-aided design (CAD) software packages instead of drawing by hand. Use the internet to find out more about these packages.

Example: How to construct a triangle if you know two sides and one angle.

Construct the triangle ABC.

- Draw a line BC 7.5 cm long.
- Draw an angle of 50° at B.
- Draw the line AB 4.1 cm long.
- Join AC to complete the triangle.

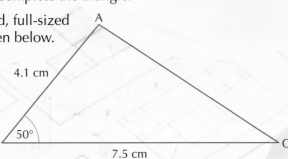

The completed, full-sized triangle is given below.

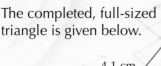

1 a Construct the triangle PQR.

   b Measure the size of ∠P and ∠R to the nearest degree.

   c Measure the length of the line PR to the nearest millimetre.

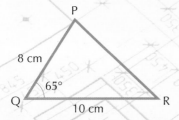

2 Construct the triangle ABC with ∠A = 100°, ∠B = 40° and AB = 8 cm.

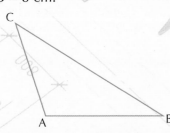

3 a Construct the trapezium ABCD.

   b Measure ∠B to the nearest degree.

   c Measure the length of the lines AB and BC to the nearest millimetre.

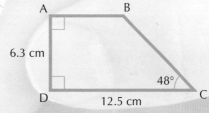

4 a Construct the quadrilateral PQRS.

   b Measure ∠P and ∠Q to the nearest degree.

   c Measure the length of the line PQ to the nearest millimetre.

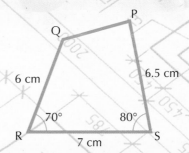

5 Construct the parallelogram ABCD with AB = 7.4 cm, AD = 6.4 cm, ∠A = 50° and ∠B = 130°.

10 Coordinates and graphs

This chapter is going to show you:

- how to recognise and draw graphs in which x or y has a fixed value
- how to recognise and draw graphs of the form $y = x$ and $y = -x$
- how to recognise and draw graphs of the form $y = x + a$ and $x + y = a$
- how to interpret and draw graphs that show real-life problems.

You should already know:

- how to use coordinates in all quadrants.

About this chapter

Graphs can save lives. If the doctor thinks you have a heart problem, you will be linked up to an electrocardiogram machine that will turn the electrical signals produced by your heart into a graph on a screen. This makes it easy to see instantly if there are any problems. It also helps to monitor very ill people, or those having operations, as any problem with their heartbeat can be seen on the graph. Graphs of all types make it easier to interpret data visually and see what is happening.

10.1 Coordinates in four quadrants

Learning objective

• To understand and use coordinates to locate points in all four quadrants

Key words

axes	coordinate
origin	quadrant
x-axis	x-coordinate
y-axis	y-coordinate

You can use **coordinates** to locate a point on a grid. The grid consists of two **axes**, called the **x-axis** and the **y-axis**. They are perpendicular (at right angles) to each other. The two axes meet at a point called the **origin**, which is numbered 0 on both axes. The four parts of the grid formed by the axes are called **quadrants**.

Coordinates are always written in pairs. The first one is the **x-coordinate** and shows how far along a point is on the x-axis. The second one is the **y-coordinate** and shows how far along the point is on the y-axis. The coordinates of the origin are (0, 0).

When you use coordinates, the first number in the pair is the x-coordinate and the second number is the y-coordinate. If you are talking in general terms about coordinates you can write any point as (x, y).

Example 1

Write down the coordinates of the points A, B, C and D on this grid.

In this grid:

• point A has the coordinates (4, 2)

• point B has the coordinates (–2, 3)

• point C has the coordinates (–3, –1)

• point D has the coordinates (1, –4).

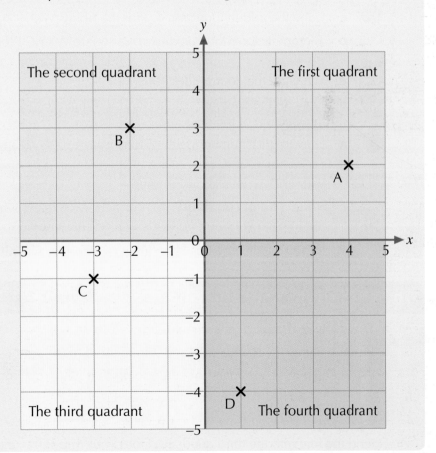

Exercise 10A

1 Look at this grid.

Write down the coordinates of the points A, B, C, D, E, F, G and H.

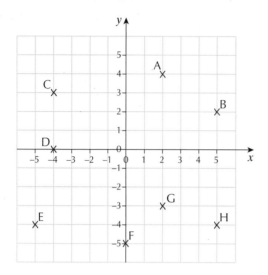

2 **a** Copy the grid in Question 1. Plot the points A(−3, −2), B(−3, 3), C(1, −2), and D(1, 3).

b Join the points, in the order given. What letter have you drawn?

3 **a** Copy the grid in Question 1. Plot the points W(2, −2), X(2, 3) and Y(−3, −2).

b The points form three vertices of a square WXYZ. Plot the point Z and draw the square.

c What are the coordinates of the point Z?

d Draw in the diagonals of the square. What are the coordinates of the point of intersection of the diagonals?

4 **a** Draw a grid, numbering the x-axis from −10 to 10 and the y-axis from −5 to 5.

b Using letters with straight sides and drawing them between coordinate points on your grid, write a short message.

c Write a list of coordinates to represent each letter, when its coordinates are joined up. In each word, separate the letters with a forward slash (/).

d Swap coordinate messages with a partner and decode each other's messages.

Investigation: Spot the rule

A **a** Draw a grid and number both axes from −8 to 8.

b Plot the points (0, 2) and (6, 8) and join them to make a straight line.

c Write down the coordinates of other points that lie on the line.

B Can you spot a rule that connects the x-coordinate and the y-coordinate?

C The rule you have found should be $y = x + 2$.

Extend the line into the third quadrant. Does your rule still work?

10.2 Graphs from relationships

Learning objective

* To draw a graph for a simple relationship

Key words

| relationship | substitute |

You should have found in the Investigation at the end of the last section that the rule for the graph there is $y = x + 2$. This shows the **relationship** between x and y.

This section shows you how to draw graphs from relationships like this. For example:

$y = x + 1$

To work out the (x, y) coordinates for the graph **substitute** any values for x to find the corresponding values of y. It is helpful to set this out in a table of values like this.

x		y
1	→	2
2	→	3
3	→	4
4	→	5
5	→	6

This gives you the (x, y) coordinate pairs you need, to draw the graph.

$(1, 2), (2, 3), (3, 4), (4, 5), (5, 6)$

You can plot these coordinates on a pair of axes, as shown on this graph.

Then you can join all the points with a straight line.

Choose any value of x beyond 5, on the line. Put this value of x into the relationship and then check from the graph that your value of y is correct.

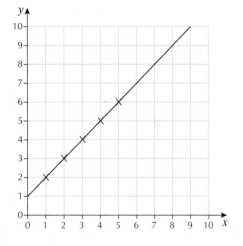

Notice that you can extend the line into the area bordered by the negative axes.

Check that the coordinates still fit the relationship.

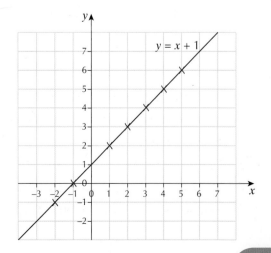

Example 2

Draw the graph of the relationship $y = 2x + 1$ for $-2 \leqslant x \leqslant 4$.

You are being asked to draw the graph for values of x from -2 to 4.

You need to calculate the value of y for the x-values of -2, -1, 0, 3 and 4.

Then use these values to draw the graph of $y = 2x + 1$ for $-2 \leqslant x \leqslant 4$.

Set up a table of values.

Hint Remember that $\leqslant$ means 'less than or equal to'.

x	$y = 2x + 1$
-2	-3
-1	-1
0	1
3	7
4	9

Write the pairs of values as coordinates.

$(-2, -3), (-1, -1), (0, 1), (3, 7), (4, 9)$

Plot these coordinates and join them in a straight line.

Always label the line.

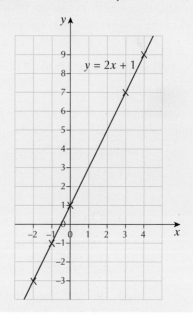

Example 3

Using x values of -4, 0 and 4, work out some coordinates (x, y) that lie on the line $y = x - 2$. Then draw the line for $-4 \leqslant x \leqslant 4$.

Substitute the x-values into the relationship and write as coordinates. This gives $(-4, -6)$, $(0, -2)$ and $(4, 2)$.

Plot these coordinates and draw the line.

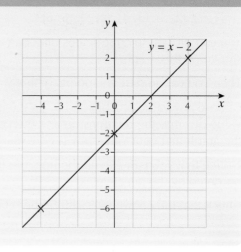

When you are asked to plot a line for a given relationship how many points do you need to be sure you are drawing it correctly?

The answer is two but always draw at least three. That way if all three points lie on the same straight line you can be almost certain that you have drawn the line correctly.

Exercise 10B

1 For each relationship:

   i complete the table of values   **ii** complete the list of coordinates

   iii plot the coordinates and draw the graphs, numbering the x-axis from -3 to 7.

   Draw all the graphs on the same set of axes.

a $y = x + 2$

x	y	Coordinates
$-1 \rightarrow$	1	$(-1, 1)$
$0 \rightarrow$	2	$(\,0, 2)$
$1 \rightarrow$		$(\,1, \quad)$
$2 \rightarrow$		$(\,2, \quad)$
$3 \rightarrow$		$(\,3, \quad)$
$4 \rightarrow$		$(\,4, \quad)$

b $y = x + 4$

x	y	Coordinates
$-1 \rightarrow$	3	$(-1, 3)$
$0 \rightarrow$	4	$(\,0, 4)$
$1 \rightarrow$		$(\,1, \quad)$
$2 \rightarrow$		$(\,2, \quad)$
$3 \rightarrow$		$(\,3, \quad)$
$4 \rightarrow$		$(\,4, \quad)$

c $y = x - 3$

x	y	Coordinates
$-1 \rightarrow$	-4	$(-1, -4)$
$0 \rightarrow$	-3	$(\,0, -3)$
$1 \rightarrow$		$(\,1, \quad)$
$2 \rightarrow$		$(\,2, \quad)$
$3 \rightarrow$		$(\,3, \quad)$
$4 \rightarrow$		$(\,4, \quad)$

   d On the same axes, draw the graph of $y = x$.

   e What do you notice about all the lines?

2 For each relationship:

   i complete the table   **ii** complete the list of coordinates

   iii plot the coordinates and draw the graphs, numbering the x-axis from -3 to 7.

   Draw all the graphs on the same set of axes.

a $y = 2x$

x	y	Coordinates
$-1 \rightarrow$	-2	$(-1, -2)$
$0 \rightarrow$	0	$(\,0, 0)$
$1 \rightarrow$		$(\,1, \quad)$
$2 \rightarrow$		$(\,2, \quad)$
$3 \rightarrow$		$(\,3, \quad)$
$4 \rightarrow$		$(\,4, \quad)$

b $y = 3x$

x	y	Coordinates
$-1 \rightarrow$	-3	$(-1, -3)$
$0 \rightarrow$	0	$(\,0, 0)$
$1 \rightarrow$		$(\,1, \quad)$
$2 \rightarrow$		$(\,2, \quad)$
$3 \rightarrow$		$(\,3, \quad)$
$4 \rightarrow$		$(\,4, \quad)$

c $y = 5x$

x	y	Coordinates
$-1 \rightarrow$	-5	$(-1, -5)$
$0 \rightarrow$	0	$(\,0, 0)$
$1 \rightarrow$		$(\,1, \quad)$
$2 \rightarrow$		$(\,2, \quad)$
$3 \rightarrow$		$(\,3, \quad)$
$4 \rightarrow$		$(\,4, \quad)$

   d On the same axes, draw the graph of $y = x$.

   e What do you notice about all the lines?

3 **i** Draw a grid, numbering the x-axis from -3 to 4 and the y-axis from -10 to 15.

   ii Use x-values of -2, 0 and 3 to work out the corresponding y-values in each relationship.

   iii Write each pair of values as a set of coordinates.

   iv Plot the coordinates and draw the graphs.

      a $y = 3x - 2$ **b** $y = 2x + 2$ **c** $y = 4x + 1$

4 Choose some of your own starting points and create a graph from each relationship.

    a $y = 2x + 3$ **b** $y = 3x + 2$ **c** $y = 2x - 3$ **d** $y = 4x - 3$

    e $y = 2x + 5$ **f** $y = 3x - 1$ **g** $y = 4x + 3$ **h** $y = 2x - 5$

 5 Draw a grid, numbering the x-axis from 0 to 5 and the y-axis from 0 to 10.

    a On your grid, draw the graph of $y = 2x + 1$.

    b What is the value of x when $y = 9$?

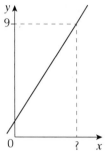

6   **a** Copy and complete this table for the relationship $y = -2x + 3$.

x	$y = -2x + 3$	Coordinates
−2	7	(−2, 7)
0		(0,)
2		(2,)

    b Draw a grid, numbering the x-axis from −2 to 2 and the y-axis from −2 to 8. Plot the coordinates and draw the graph of $y = -2x + 3$.

    c Find some coordinates to help you draw the graphs of these relationships.

      i $y = -x - 1$ **ii** $y = -3x + 1$

7 Choose some of your own starting points and create a graph for each relationship.

    a $y = -2x + 1$ **b** $y = -3x + 2$ **c** $y = -2x - 1$ **d** $y = -4x$

    e $y = -2x$ **f** $y = -3x - 1$ **g** $y = -4x - 3$ **h** $y = -2x + 2$

Investigation: $y = x^2$

A Draw a graph for the relationship $y = x^2$. Number the x-axis from −5 to 5.

B Use your graph to help you to find the value of $\sqrt{19}$, without using your calculator.

C Check how accurate you were by working out $\sqrt{19}$ on your calculator.

D Describe the difference between the graph of $y = x^2$ and all the other graphs you have drawn in this section.

10.3 Predicting graphs from relationships

Learning objective

- To understand the connection between pairs of coordinates and the relationship shown on a graph

You do not always need to draw a graph to work out whether points, or pairs of coordinates, will be on it. You can use the known relationship.

Example 4

Are the points (2, 5) and (3, 9) on the graph of $y = 2x + 3$?

Look at the coordinates (2, 5).

When $x = 2$:

$$y = 2x + 3$$
$$= 2 \times 2 + 3$$
$$= 7 \text{ (which is not 5)}$$

So, the point (2, 5) is not on the line $y = 2x + 3$.

Look at the coordinates (3, 9).

When $x = 3$:

$$y = 2x + 3$$
$$= 2 \times 3 + 3$$
$$= 9$$

So, the point (3, 9) is on the line $y = 2x + 3$.

You can check this by looking at the graph of $y = 2x + 3$.

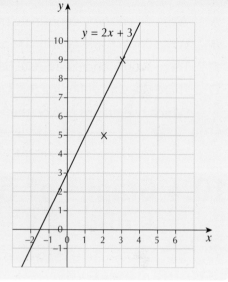

Exercise 10C

Answer the first five questions without drawing graphs.

1
  a Is the point (3, 5) on the graph of $y = x + 2$?
  b Is the point (4, 2) on the graph of $y = x - 2$?
  c Is the point (−3, 3) on the graph of $y = x + 3$?
  d Is the point (2, 10) on the graph of $y = 5x$?
  e Is the point (3, 8) on the graph of $y = 3x$?
  f Is the point (−2, −8) on the graph of $y = 4x$?

2 Which of these lines does the point (2, 3) lie on?

$y = x - 1$ $y = x + 1$ $y = 2x$ $y = 3$

3 Which of these lines does the point (3, 6) lie on?

$y = x + 2$ $y = x + 3$ $y = 2x$ $y = 3$

4 Write down two relationships with graphs that will pass through the point (1, 3).

5 Write down two relationships with graphs that will pass through the point (2, 8).

6 Draw a grid, numbering the *x*-axis from 0 to 3 and the *y*-axis from 0 to 7.

By drawing the two graphs, find the coordinates of the point where the graphs of $y = x + 1$ and $y = 2x$ intersect (cross each other).

PS **7** Work out the area of the shape enclosed by the graphs of $y = x + 3$, $y = x - 3$, $x + y = 3$ and $x + y = 9$.

PS **8** What is the special name of the quadrilateral formed by these four lines?

$$y = 2x + 2 \quad y = -2x + 2 \quad y = \frac{1}{2}x - 1 \quad y = -\frac{1}{2}x + 5$$

Challenge: Common coordinates

A Write down some coordinate pairs that fit the relationship $y = 4x - 1$.

B Write down some coordinate pairs that fit the relationship $y = 2x + 2$.

C There is only one coordinate pair that fits both relationships. Which is it?

D There is a way of finding this one coordinate pair by drawing graphs.

Which two graphs do you think you would need to draw? Check to find out.

10.4 Graphs of fixed values of *x* and *y*, *y* = *x* and *y* = −*x*

Learning objectives

• To recognise and draw line graphs with fixed values of *x* and *y*

• To recognise and draw graphs of $y = x$ and $y = -x$

Graphs of fixed values of *x* and *y*

What do you notice about this set of coordinates?

(0, 4), (1, 4), (2, 4), (3, 4), (4, 4)

The second number, the *y*-coordinate, is always 4.

In other words, $y = 4$.

Look what happens when you plot the graph.

This is the graph of the relationship $y = 4$.

Whenever the value of *y* is a fixed number, the graph of the relationship will be a horizontal line.

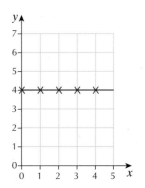

When you repeat this investigation for a fixed x-value, such as $x = -3$, you find that the graph of the relationship is always a vertical line, as shown.

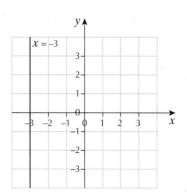

The graph of $y = x$

Look at the line on this graph.

The coordinates of some of the points on the line are:

$(-3, -3)$, $(-2, -2)$ $(-1, -1)$, $(0, 0)$, $(1, 1)$, $(2, 2)$, $(3, 3)$

You should notice that the x-coordinate is the same as the y-coordinate in each case.

In other words, the relationship is $y = x$.

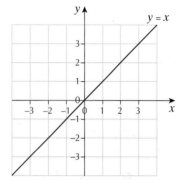

The graph of $y = -x$

The coordinates of the line on this grid are:

$(-3, 3)$, $(-2, 2)$, $(-1, 1)$, $(0, 0)$, $(1, -1)$, $(2, -2)$, $(3, -3)$

In this case the y-coordinate is the negative value of the x-coordinate.

If x is positive, y is negative. If x is negative, y is positive as two negatives make a positive.

This is the line for the relationship $y = -x$.

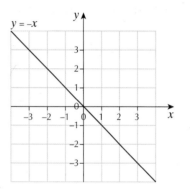

Exercise 10D

1 Write down the name of the straight line that goes through each pair of points, on the diagram.

    a A and B   **b** C and A   **c** E and H

    d G and C   **e** L and J   **f** K and H

    g D and F   **h** D and E   **i** G and I

    j A and K

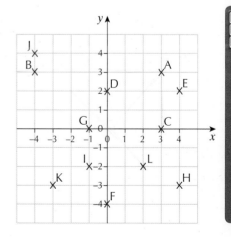

2 Draw a grid, numbering both axes from −5 to 5.

Draw each of these graphs on the same grid. Remember to label them.

a $y = -1$ **b** $y = -4$ **c** $y = x$ **d** $x = -1$

e $x = 3$ **f** $x = -2\frac{1}{2}$ **g** $y = 4$ **h** $y = -x$

3 Write down the letters that are on each line.

a $x = -2$ **b** $y = 1$ **c** $y = -3$

d $y = x$ **e** $x = 3$ **f** $x = -4$

g $y = -1$ **h** $x = -1$ **i** $y = -2$

j $y = -x$

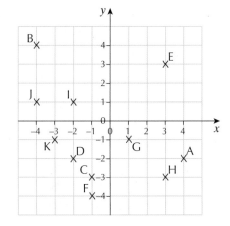

4 **a** Draw a suitable set of coordinate axes.

Draw each pair of lines on the same grid. Write down the coordinates of the point where they cross.

i $x = -1$ and $y = 3$ **ii** $x = 4$ and $y = -1$ **iii** $y = -5$ and $x = -6$

b Write down the coordinates of the point where each pair of lines cross. Do not draw them.

i $x = -9$ and $y = 5$ **ii** $x = 28$ and $y = -15$ **iii** $y = -23$ and $x = -48$

5 These are the coordinates of 12 points.

A(−2, −3)   B(−3, −5)   C(7, −3)   D(−2, −5)   E(−2, −2)

F(6, −6)   G(3, 7)   H(7, −4)   I(−3, −4)   J(7, 7)

K(−5, 5)   L(−3, −6)

Write down the relationship for the straight line that passes through both points, in each of the following pairs.

Then draw a suitable set of axes and plot the points on the graphs, to check your answers.

a A and C   **b** B and D   **c** C and H   **d** A and D   **e** E and J

f F and L   **g** B and I   **h** C and J   **i** F and K   **j** H and I

The lines $x = 2$, $y = 7$ and $y = x$ are drawn on this grid.

They enclose a triangle with an area of
$\frac{1}{2} \times 5 \times 5 = 12\frac{1}{2}$ square units.

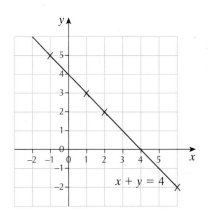

A Draw at least two more sets of lines, one with a fixed value of x, one with a fixed value of y and $y = x$.

B Work out the area of the triangle they enclose each time.

C Write down a rule for working out the area, without drawing the lines.

D Use your rule to work out the areas of the triangles enclosed by these sets of lines, without drawing.

    a $x = 1$, $y = 8$ and $y = x$ **b** $x = -2$, $y = 3$ and $y = x$

    c $x = -5$, $y = 5$ and $y = x$ **d** $x = -7$, $y = 1$ and $y = -x$

    e $x = -5$, $y = -2$ and $y = -x$

10.5 Graphs of the form $x + y = a$

Learning objective

- To recognise and draw graphs of the form $x + y = a$

This table shows pairs of values for x and y. Note that the numbers in each pair add up to 4.

x	y	Total	Coordinates
1	3	4	(1, 3)
2	2	4	(2, 2)
−1	5	4	(−1, 5)
6	−2	4	(6, −2)

You can plot the coordinates on a graph and join them with a straight line.

Because the x and y values add up to 4, the relationship for this line is $x + y = 4$.

This general relationship is referred to as $x + y = a$ where a is a constant number.

Notice that the line $x + y = 4$ goes through (0, 4) on the y-axis and (4, 0) on the x-axis.

All lines of the form $x + y = a$ will pass through the points (0, a) and (a, 0).

Exercise 10E

1 Draw up a table of values to show ways in which x and y can add up to 9.
Remember to include some negative values.
Draw a grid, numbering both axes from −3 to 10.
Plot the coordinates on a grid and join up the points.
Label the line $x + y = 9$.

2 Draw a grid, numbering both axes from −6 to 10.
Draw up a table of values and use it to draw the line for each relationship.

    a $x + y = 3$ **b** $x + y = 6$ **c** $x + y = -2$

(MR) 3 Explain why the lines $x + y = 2$ and $x + y = 7$ will never meet.

(PS) 4 Which of these lines is the same as $y = x$?
There may be more than one answer.

    a $x = y$ **b** $y + x = 0$ **c** $x = -y$ **d** $y - x = 0$ **e** $y = x + 0$

5 There are 16 points marked on this grid.
Write down the relationship for the line that passes through both points, in each pair.

    a Q and H   **b** B and D
    c A and J   **d** L and N
    e C and P   **f** E and F
    g K and M   **h** G and I

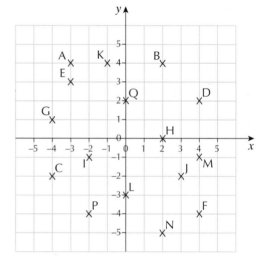

(PS) 6 **a** Draw a grid, numbering both axes from −2 to 6.
      On the grid, draw the lines $x + y = 5$ and $y = x$.

    b Where do the lines intersect?

    c What is special about the coordinates of the point where they intersect?

    d Write down a rule for working out the intersection point of a line of the form
      $x + y = a$ and the line $y = x$.

    e Without drawing a graph, work out the coordinates of the point where each pair
      of lines intersect.

      i $x + y = 4$ and $y = x$ **ii** $x + y = 11$ and $y = x$ **iii** $x + y = -8$ and $y = x$

10.6 Graphs from the real world

Learning objectives

- To learn how graphs can be used to represent real-life situations

- To draw and use real-life graphs

When you fill your car with petrol, both the amount of petrol you've taken and its cost are displayed on the pump. The price of petrol varies but, in this example, the cost is taken as being about £1.40 for one litre.

The table shows the costs of different quantities of petrol, as displayed on a petrol pump.

Petrol (litres)	5	10	15	20	25	30
Cost (£)	7	14	21	28	35	42

This information can also be represented by these pairs of coordinates.

(5, 7) (10, 14) (15, 21) (20, 28) (25, 35) (30, 42)

These pairs have been plotted on the grid, to give a graph that relates the cost of petrol to the quantity bought.

This is an example of a **conversion graph**. You can use it to find the cost of any quantity of petrol, or to find how much petrol you can buy for a given amount of money.

Conversion graphs are usually straight-line graphs.

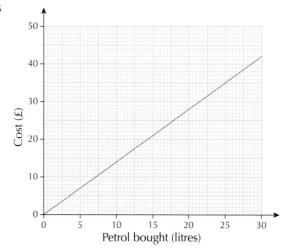

Example 5

The graph shows how much a fast food shop charges to deliver pizzas.

The vertical axis shows the charge and the horizontal axis shows the number of pizzas.

a How much does the shop charge to deliver four pizzas?

b The shop has a fixed delivery charge. How much is it?

c How much does the shop charge for each pizza?

d Why is the graph drawn as a dashed line?

e Use the graph to help write a rule that can be used to calculate what the shop charges to deliver pizzas, where d is the total and n is the number of pizzas delivered.

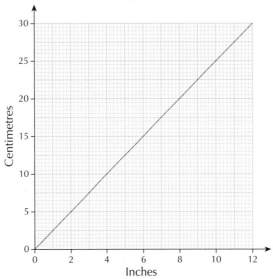

a £30

b The delivery charge is £2, as this is the cost marked for zero pizzas.

c £7

d Because you cannot include fractions of a pizza. The shop only delivers whole pizzas so the line just shows a trend, rather than a set of values.

e $d = 2 + 7n$

Exercise 10F

1 This is a conversion graph between centimetres (cm) and inches.

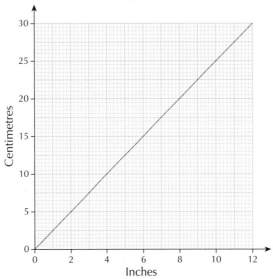

a Express each length in centimetres.

   i 3 inches   **ii** 9 inches

b Express each length in inches.

   i 2 cm   **ii** 12 cm   **iii** 25 cm

2 a Copy and complete the following table for the exchange rate between the US dollar and the British pound.

Dollar ($)	5	10	15	20	25
Pounds (£)			9.45	12.60	15.75

b Use the data from this table to draw a conversion graph between pounds and dollars.

c Use your graph to convert each amount to pounds.

   i $2   **ii** $18   **iii** $22.50

d Use your graph to convert each amount to dollars.

   i £5   **ii** £12   **iii** £14.80

(MR) **3** The mass of a crate is 7 kg. Packets of cereal, each of mass 175 g, are packed into it.

a Draw a graph to show the total mass of the crate plus the packets of cereals packed into it, against the number of packets of cereal in the crate.

b Find, from the graph, the number of packets of cereal that can be packed in the crate, to give a total mass as close to 20 kg as possible.

4 The conversion graph shows the relationship between dollars ($) and pounds (£).

a How many pounds would you get for:

   i $20   **ii** $28?

b How many dollars would you get for:

   i £40   **ii** £10?

c Write down a formula connecting the number of dollars (D) and the number of pounds (P) in the form $D = aP$, where a is a number.

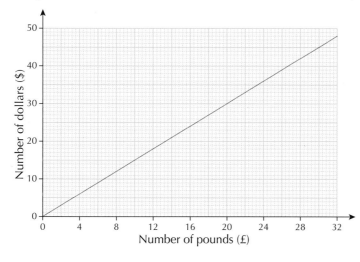

5 This graph shows the distance travelled by an aeroplane during an interval of five hours.

   a How far has the aeroplane travelled in five hours?

   b How long does it take the aeroplane to travel 270 km?

   c Write down a formula for the distance (D) travelled by the plane in H hours.

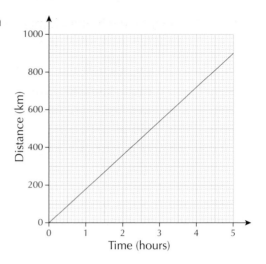

6 This graph shows the amount of fuel in the tank of the aeroplane in question **5** as it travels for the five hours.

   a How much fuel is there left in the aeroplane's tank after two hours?

   b How much fuel has the aeroplane used in the first five hours?

   c Write down a rule that can be used to find the amount of fuel (F) in the tank of the aeroplane after H hours, in the form $F = 2500 - aH$, where a is a number.

   d The airlines regulations say that the amount of fuel should never fall below 100 gallons. How much further can the aeroplane fly before it must land? You will need to use the graph in question 5 to help you answer this question.

 7 Two stations, A and B, are 100 km apart. At 12:00 noon, train 1 leaves station A and travels at an average speed of 80 km/h to station B. At the same time train 2 leaves station B and travels at an average speed of 100 km/h to station A.

This sketch graph shows the journeys of both trains.

   a Draw an accurate graph to work out how far from station A the trains are when they pass each other.

   b Work out what time it is when they pass each other.

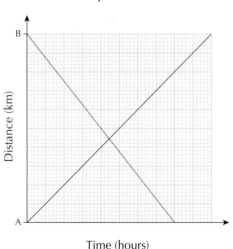

MR **8** Ivor Wrench, the plumber, uses the formula $C = 30 + 25H$ to work out how much to charge for a job, where C is the charge and H is the time, in hours, the job takes.

Walter Pipe, another plumber, uses the formula $C = 40 + 20H$.

a Copy this set of axes and draw a graph to show both formulae.

b For what length of job do the two plumbers charge the same amount?

c An engineer estimates that a plumbing job will take between 1 and 4 hours. Which plumber would be the cheaper? Explain your choice.

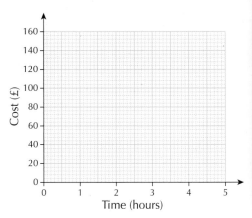

Challenge: Taxi fares

A A taxi company called Streetcars charges a basic charge of £2.60 plus £1 per kilometre for a taxi journey.

City Taxis, their rival company, charge a straight £1.50 per kilometre.

Draw graphs, on the same grid, to show how much each firm charges for journeys up to 10 km.

B Jack travels from his home in Corbey to work in Melchester.

He hires a taxi from one of these two companies to take him to a local station at Midstone or Darly, then takes the train to Melchester.

Midstone station is two kilometres from his home. The train fare to Melchester from Midstone is £28.

Darly station is 10 kilometres from his home. The fare from Darly station to Melchester is £18.

What is the cheapest way for Jack to travel to Melchester? Show your working.

Ready to progress?

I can plot coordinates in all quadrants.
I can use a table of values to work out coordinates and plot them.
I can read values from conversion graphs.

I can recognise and draw lines such as $x = 3$ and $y = -1$ in all four quadrants.
I can recognise and draw lines of relationships of the form $y = x$, $y = x + a$ and $y = -x$.
I can recognise and draw lines of relationships of the form $x + y = a$.
I can read values from real-life graphs with a fixed rate plus a variable rate.

Review questions

1 a Write down the next two numbers in this sequence.

   1, 4, 7, 10, 13, …, …

b This is a sequence of coordinates.

   (1, 2), (4, 6), (7, 10), (10, 14), (13, 18)

   Write down the next two coordinates in the sequence.

c Work out the 25th coordinate in the sequence.

2 This graph is a straight line showing the relationship $y = 2x - 3$.

a A point on the line $y = 2x - 3$ has an x-coordinate of 40.

   What is the y-coordinate of this point?

b A point on the line $y = 2x - 3$ has a y-coordinate of 37.

   What is the x-coordinate of this point?

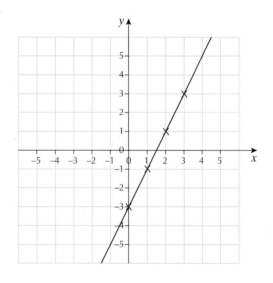

PS **3** The fractions $\frac{3}{5}$ and $\frac{5}{7}$ are plotted as coordinates.

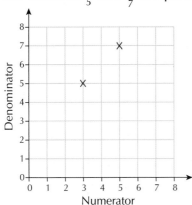

a Join the points and extend the straight line.

b Write down one other fraction that lies on this line.

c What is the relationship for the line?

d Which of these fractions lie on the line?

   i $\dfrac{10}{12}$ **ii** $\dfrac{23}{24}$ **iii** $\dfrac{100}{98}$ **iv** $\dfrac{99}{101}$

e Explain why all fractions that are plotted below the line $y = x$ will be improper.

 4 Will sets off on a cycle trail at 9:00 am, at a steady speed of 15 km/h.

An hour later Jill, his wife, sets off on the same trail, at a steady speed of 20 km/h.

The sketch graph shows the distances they travelled and their journey times.

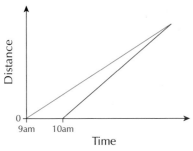

a Draw an accurate graph to show their correct distances and times.

Immediately Jill catches up with Will, they turn round and cycle back to the start of the trail together, at a steady speed of 12 km/h.

b What time do they get back to the start? Show all your working.

Challenge
Travelling abroad

Touring Europe

When you visit Europe, you will discover that its countries measure distances in kilometres instead of miles. The distances shown on road signs can therefore be confusing.

It helps if you have ready a converter between miles and kilometres, like the one in this graph.

You can use this graph to convert any distance in either miles or kilometres into the other.

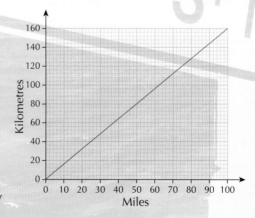

A Kath's trip to the USA

Kath took her family on a driving tour of the USA.

She found it was difficult to work out the cost of petrol because the USA gallon is different from the UK gallon.

Kath's son David managed to find out two facts about US gallons and UK gallons.

- 10 UK gallons are equivalent to 45.46 litres.

- 10 US gallons are equivalent to 37.85 litres.

Kath's daughter Helen said she needed a conversion graph from gallons to litres for UK gallons, and another one for US gallons.

1 Draw the two conversion graphs, taking the scales up to ten gallons and up to 50 litres.

2 Use your graphs to find out how many gallons of each type are equivalent to 30 litres.

3 Use your answer to question **2** to draw a conversion graph for US gallons and UK gallons.

4 Use your graph to find out:

   a how many US gallons are equivalent to 8 UK gallons

   b how many UK gallons are equivalent to 5 US gallons

5 As they left the UK, Grandad said the price of petrol was £6.36 per gallon.

   What would be the equivalent cost in pounds for a US gallon?

B Pete's trip to Europe

Pete took his family touring in Europe.

These are some of the questions his children asked on the long drives.

What did he answer?

1 That signpost said Paris is 80 km away, how many miles is that?

2 It feels like we've driven 200 miles! How many kilometres is that?

3 Mum says it's 200 kilometres to our next stop, how many miles is that?

4 Bruges is 150 kilometres away. If we drive at 70 miles per hour, how long will it take us to get there?

5 How many kilometres are equivalent to 1 mile?

Hint Pete answered these questions to one decimal place.

11

Percentages

This chapter is going to show you:

- how to interpret percentages as fractions or decimals
- how to use percentages greater than 100%
- how to work out a fraction or a percentage of a quantity, with and without a calculator
- how to work out the result of a percentage change
- how to solve problems involving percentage changes.

You should already know:

- equivalences between simple fractions, decimals and percentages
- how to simplify fractions
- how to calculate a simple percentage of a whole number.

About this chapter

Few people can resist a good sale. The percentages taken off prices tell you instantly how much they have been reduced and help you judge straight away if you might be getting a good bargain. A reduction of £5 is a good deal when the original price was £10 (50% off), but it's less attractive when the starting price is £500 – a reduction of just 1%! Becoming comfortable working with percentages will help you to become a better bargain hunter, and enable you to compare lots of other kinds of important data as well such as the interest you will get on bank savings accounts.

11.1 Fractions, decimals and percentages

Learning objectives

- To understand the equivalence between a fraction, a decimal and a percentage
- To understand and use percentages greater than 100%

Key words	
decimal	fraction
per cent (%)	percentage

Fractions, **decimals** and **percentages** can all be used to compare quantities or measurements.

Look at this pie chart, which shows how a class voted for their favourite of three sports.

It shows that:

$$\frac{1}{4} + \frac{1}{3} + \frac{5}{12} = 1$$

$\frac{1}{4}$ is 25% (25 **per cent**) or 0.25 of the circle.

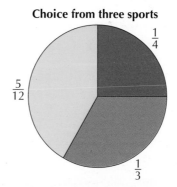

Choice from three sports

Example 1

What percentages are $\frac{1}{3}$ and $\frac{5}{12}$?

$$\frac{1}{3} = \frac{1}{3} \times 100\%$$
$$= 33\frac{1}{3}\%$$

This is a common percentage that you should know.

For a less common fraction, such as $\frac{5}{12}$, first convert it to a decimal.

$$\frac{5}{12} = 0.4166\ldots$$

Then multiply the decimal by 100 to change it to a percentage.

$$\frac{5}{12} = 41.7\%, \text{ rounding the answer to one decimal place (1dp).}$$

Example 2

In a certain town there are 47 sunny days in July and August.

What percentage of all the days in the two months is that?

There are 62 days in July and August.

The fraction of sunny days is $\frac{47}{62}$.

$$\frac{47}{62} = 47 \div 62$$
$$= 0.7580\ldots$$

So $\frac{47}{62} = 75.8\%$.

You can do this in one calculation.

$$47 \div 62 \times 100 = 75.8\%$$

Percentages can be more than 100%.

100% = 1 so, for example, 200% = 2 and 150% = 1.5.

Example 3

These are two copper pipes.

Pipe X   240 mm

Pipe Y   600 mm

a What percentage of the length of pipe Y is the length of pipe X?

b What percentage of the length of pipe X is the length of pipe Y?

    a Pipe X is $\frac{240}{600} = \frac{2}{5}$ of pipe Y. $\frac{2}{5} = 40\%$

    b Pipe Y is $\frac{600}{240} = \frac{5}{2}$ of pipe X. $\frac{5}{2} = 2.5 = 250\%$

This means that pipe Y is 2.5 or $2\frac{1}{2}$ times as long as pipe X.

Exercise 11A

 1   Change these fractions to percentages.

    a $\frac{4}{5}$ **b** $\frac{1}{8}$ **c** $\frac{7}{20}$ **d** $\frac{5}{4}$ **e** $\frac{7}{4}$

 2   Write each of these decimal numbers as a percentage.

    a 0.7   **b** 0.07   **c** 1.7   **d** 1.07   **e** 1.77

 3   Write these percentages as fractions. Give each answer as simply as possible.

    a 80%   **b** 35%   **c** 2%   **d** 120%   **e** 225%

4   Copy and complete this table.

The first row has been done for you.

Percentage	Fraction	Decimal
40%	$\frac{2}{5}$	0.4
95%		
		1.9
4%		
	$2\frac{3}{5}$	

5   **a** Write $\frac{2}{3}$ as a percentage.   **b** Write $\frac{5}{3}$ as a percentage   **c** Write $\frac{8}{3}$ as a percentage.

6   There are 83 cars in a car park.

38 are more than 10 years old and five of those are more than 20 years old.

    a Work out the percentage of the cars that are more than 10 years old.

    b Out of the group of cars that are more than 10 years old, what percentage are more than 20 years old?

(FS) **7** In 2000 a flat was sold for £125 000.

In 2010 the same flat was sold for £192 000.

    a What percentage of the 2010 price is the 2000 price?

    b What percentage of the 2000 price is the 2010 price?

(PS) **8** Visitors to a website are asked where they live.

The answers are shown in this pie chart.

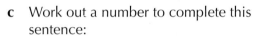

Where do you live?

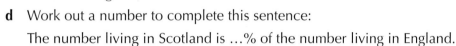

England, 32

Wales, 57

Scotland, 45

    a What percentage live in each country?

    b If you just look at the people who live in England or Wales, what percentage live in Wales?

    c Work out a number to complete this sentence:

       The number living in England is …% of the number living in Scotland.

    d Work out a number to complete this sentence:

       The number living in Scotland is …% of the number living in England.

(MR) **9** **a** Write each of these fractions as a percentage.

       $\frac{3}{8}$ $\frac{7}{8}$ $\frac{11}{8}$ $\frac{15}{8}$

    b How will the sequence continue?

10 Rewrite each of these statements as an addition of percentages.

    a $\frac{2}{5} + \frac{3}{8} = \frac{31}{40}$ **b** $\frac{5}{6} + \frac{2}{3} = 1\frac{1}{2}$

11 $80\% + 45\% = 125\%$

    a Rewrite this as an addition of decimals.

    b Rewrite this as an addition of fractions.

Investigation

Here is the start of a sequence of numbers.

  1 1 2 3 5 8 13 21 34 …

A Write down the next three numbers in the sequence.

B You can present each number as a percentage of the previous number.

   For example:

   $\frac{1}{1} = 100\%$ $\frac{2}{1} = 200\%$ $\frac{3}{2} = 150\%$

   Continue this for five or more further pairs of numbers in the sequence.

C What do you notice about your percentages?

11.2 Fractions of a quantity

Learning objective

• To work out a fraction of a quantity without using a calculator

You can work out fractions by a mixture of multiplication and division.

Example 4

There are 360 animals on a farm.

$\frac{1}{5}$ are cows and $\frac{3}{8}$ are sheep.

a Work out the number of cows.

b Work out the number of sheep.

    a To work out $\frac{1}{5}$ divide 360 by 5. $360 \div 5 = 72$ There are 72 cows.

    b To work out $\frac{3}{8}$ first find $\frac{1}{8}$ by dividing 360 by 8. $360 \div 8 = 45$

    Then $\frac{3}{8}$ of $360 = 45 \times 3 = 135$.   There are 135 sheep.

Exercise 11B

(PS) 1 A marathon is just over 26 miles.

Janice says: 'When I have run 20 miles I shall have completed more than four-fifths of the marathon.' Is this correct? Give a reason for your answer.

2 Work out:

    a $\frac{1}{3}$ of £69   **b** $\frac{2}{3}$ of £69   **c** $\frac{4}{3}$ of £69.

3 Work out:

    a $\frac{1}{10}$ of £637   **b** $\frac{7}{10}$ of £637   **c** $\frac{11}{10}$ of £600   **d** $\frac{17}{10}$ of £600.

4 Work out:

    a $\frac{2}{3}$ of 120 cm   **b** $\frac{3}{5}$ of 120 cm   **c** $\frac{5}{8}$ of 120 cm   **d** $\frac{8}{15}$ of 120 cm.

(PS) 5 Look at this statement.

$\frac{2}{3}$ of 90 is the same as $\frac{3}{4}$ of N.

What is the value of N?

(PS) 6 a Work out each amount and then arrange the amounts in two separate groups.

    i $\frac{2}{3}$ of 60   **ii** $\frac{3}{8}$ of 128   **iii** $\frac{5}{8}$ of 64   **iv** $\frac{5}{3}$ of 24   **v** $\frac{3}{2}$ of 32   **vi** $\frac{2}{3}$ of 72

    b Work out one more fraction of a quantity for each group.

7 Work out each of these fractions of one hour. Give your answers in minutes.

    a $\frac{3}{4}$ **b** $\frac{4}{3}$ **c** $\frac{4}{5}$ **d** $\frac{5}{4}$ **e** $\frac{5}{6}$ **f** $\frac{6}{5}$

(PS) 8 Zeta has £800.

She spends three-quarters of it on clothes.

She spends three-quarters of what is left on food.

She spends three quarters of what is left on a hair-do.

How much does she have now?

9 Work out each amount, correct to the nearest penny.

a $\frac{2}{3}$ of £10   **b** $\frac{3}{10}$ of £7   **c** $\frac{3}{7}$ of £10   **d** $\frac{3}{5}$ of £12   **e** $\frac{5}{8}$ of £22

(PS) 10 Here is a very old story.

A man leaves 17 camels in his will.

He leaves one-half of his camels to his oldest son, one-third to his second son and one-ninth to his youngest son.

When he dies the sons do not know how to share out the camels because those fractions of 17 are not whole numbers.

A friend says: 'Don't worry, I have the solution. I shall give you one camel so now you have 18.

'The oldest son can have one-half of 18, which is 9.

'The second son can have one-third of 18, which is 6.

'The youngest son can have one-ninth of 18, which is 2.

'9 + 6 + 2 = 17 so there is one camel left and I can take back my camel.'

Show that the instructions in the will have not been carried out correctly. Explain why there is one camel left over.

Challenge: Filling in fractions

A Show that $\frac{3}{2}$ of 40 = $\frac{2}{3}$ of 90.

B In part **A** you showed that 40 and 90 can make this statement correct:

$\frac{3}{2}$ of … = $\frac{2}{3}$ of …

Find another pair of whole numbers to make this statement correct.

C Can you find any more pairs of whole numbers to make the statement in part **B** correct?

11.3 Calculating simple percentages

Learning objective

• To work out a percentage of a quantity, without using a calculator

Key words	
deposit	interest

You have learnt how to work out a fraction of a quantity. You can work out a percentage of a quantity in a similar way.

Example 5

A woman earns £870 in one week.

a She pays 12% into a pension fund. How much is that?

b After other stoppages she takes home 72% of her earnings. Work out her take-home pay.

If you do not have a calculator this is how you can work out the answers.

a 10% of £870 is £87   This is one-tenth of £870.

1% of £870 is £8.70   This is one-hundredth of £870.

12% is £87 + (£8.70 × 2) = £104.40

b 72 is 12 × 6

So 72% of £870 = £104.40 × 6 = £626.40.

There are other ways to work out these percentages.

You could write 12% as a fraction and work out $\frac{12}{100}$ of £870.

You could say 72% = 50% + 20% + 2% and add those percentages.

Example 6

In 1980 the value of a painting was £6000.

In 2010 the value was 240% of its value in 1980.

Work out the value in 2010.

240% = 2.4

2.4 × 6000 = 14 400   The value is £14 400.

Remember that percentages can be more than 100%.

In this case it means that for every £100 it was worth in 1980 it is worth £240 in 2010.

Exercise 11C

1 Work out 11% of:

   a £700   **b** £40   **c** £8   **d** £6.25   **e** £32 000   **f** £14 000 000.

2 Work out:

   a 9% of £300   **b** 19% of £400   **c** 32% of £2000   **d** 91% of £3000.

FS **3** The cost of a holiday for a couple is £1800.

They must pay a **deposit** of 20%.

How much is left to pay after they have paid the deposit?

4 Adele buys some furniture for £4500.

She pays a deposit of 20%.

She pays the remainder in 24 monthly payments.

How much is each monthly payment?

5 Work out:

a   50% of 31 kg   b   25% of 45 m   c   120% of 19 litres   d   15% of 72 cm

e   5% of 1.5 kg   f   2% of £20 000   g   125% of 4800 people   h   30% of 420 m.

6 Work out:

a   20% of 23 kg   b   30 % of 23 kg   c   120% of 23 kg   d   105% of 23 kg.

7 Work out:

a   25% of £42.00   b   2.5% of £42.00   c   1% of £42.00   d   1.5% of £42.00.

8 Work out:

a   12.5% of 72 g   b $33\frac{1}{3}$% of 90 cm   c   87.5% of £5.60   d $66\frac{2}{3}$% of 39 cm.

9 18% of £X is £11.25.

Use this fact to work out:

a   36% of £X b   54% of £X c   6% of £X d   180% of £X.

10 29% of £69.00 is £20.01.

Use this fact to work out:

a   129% of £69.00   b   71% of £69.00.

11 a   Show that 75% of £20 is the same as 20% of £75.

b   Show that 36% of £90 is the same as 90% of £36.

c   Tom's teacher set this problem.

Work out 83% of £50.

It is easier to work out 50% of £83 and it has the same answer.

Explain why Tom is correct.

Financial skills: Credit cards

Many people buy goods with their credit cards.

When they do, they get a bill at the end of the month.

They must pay a minimum amount. **Interest** is charged on the remaining amount they owe. This is a fee for borrowing the money.

Elaine has a credit card. The balance at the end of June is £320.

She decides to pay £50.

The rate of interest is 1% per month.

A What is the amount remaining to pay after she has paid £50?

B How much interest will she have to pay?

C How much will she owe at the end of July if she buys nothing else with the credit card?

D Copy this table and record your answers so far.

Owing at the end of June	£320.00
Payment made	£50.00
Balance due	
Interest charged	
Amount owing at the end of July	

E At the end of October Elaine's credit card bill was £450.

She paid £100 and was charged interest of 1% per month on the remainder.

During November she spent another £185 on her credit card.

Copy and complete this table.

Owing at the end of October	£450.00
Payment made	
Balance due	
Interest charged	
Spent in November	
Owing at the end of November	

11.4 Percentages with a calculator

Learning objectives

• To use a calculator to work out a percentage of a quantity

• To know when it is appropriate to use a calculator

Sometimes you can work out a percentage easily without a calculator.

If the calculation is complicated it is more efficient to use a calculator.

4285 people vote in an election. 32.3% vote for Ms White.

How many people vote for Ms White?

32.3% as a decimal is 0.323.   This is $32.3 \div 100$.

$0.323 \times 4285 = 1384.05....$

1384 people vote for Ms White.

In this example you should round the answer to a whole number.

Some calculators have a percentage button. If yours does, make sure you know how to use it correctly.

You can always calculate a percentage by converting it to a decimal first.

Exercise 11D

1 Work out these percentages without a calculator and then check your answer with a calculator.

    a   25% of 72 m   **b**   60% of 350 km   **c**   15% of 40 000 spectators

2 Put these in order of magnitude, smallest first.

    a   72% of 741   **b**   84% of 625   **c**   96% of 555

(MR) 3 **a** Work out:

    i   37% of 4321 people   **ii**   61.7% of 4321 people   **iii**   1.3% of 4321 people.

    b   Show that the three answers in part a add up to 4321. Explain why.

4 In an election 18 319 people voted.

    The Reds gained 23.5% of the votes. The Whites gained 33.9%.
    The Blues gained 9.1%.

    How many votes did each party gain?

5 A politician is talking to a meeting of 324 people.

    He says: 'I know that 99% of the people in this room agree with me.'

    How many people is that?

(FS) 6 A concert hall charges an extra 1.5% if you pay for tickets with a credit card. Jason buys three tickets for £67.50 each. How much will he be charged, in total, for paying with a credit card?

(FS) 7 Tickets for a festival cost £215 each and there is a booking fee of 2.5%.

    Work out the total cost of four tickets.

8 Work out:

    a   43% of £285   **b**   4.3% of £285   **c**   143% of £285.

(MR) 9 **a** Work out:

    i   54% of £365   **ii**   74% of £365   **iii**   94% of £365   **iv**   114% of £365.

    b   The questions and answers in part **a** form a sequence.

    What is the next term in the sequence?

10 This table shows the populations of four cities in 1861.

London	Birmingham	Sheffield	Preston
2 804 000	296 000	185 000	83 000

a The population of London in 1991 was 238% of the population in 1861.

Show that the London population in 1991 was about 6 674 000.

b The corresponding percentages for Birmingham, Sheffield and Preston were 351%, 241% and 169%. Work out the populations of those cities in 1991.

11 In the 2010 UK General Election, 29 687 604 people voted.

This table shows the percentages that voted for the three main parties.

Conservative	36.1%
Labour	29.0%
Liberal Democrat	23.0%

a Work out how many people voted for each party. Give your answers to the nearest million.

b To the nearest million, how many of the people who voted did not vote for one of these parties?

12 The members of a club are voting about a change in the rules.

65% must be in favour for the change to take place.

The club has 823 members.

530 vote for a change in the rules. Is that enough for the change to take place? Give a reason for your answer.

Financial skills: Interest on savings

When you put money in a savings account in a bank or building society, you can be paid money for doing so. This is called the interest. It is a percentage of the amount you have in the account. It will be paid per year.

A Gareth has £4750 in a savings account. The rate of interest is 2.7%.

  a How much interest will he earn after one year?

  b What is the equivalent to this as a weekly amount?

  c If the interest rate is reduced to 2.3%, how much less will Gareth's money earn over one year?

B Copy and complete this table to show how much different amounts will earn over a year for different rates of interest.

Amount	0.5 % interest	1% interest	2% interest	3.5% interest
£80			£1.28	
£250		£2.25		
£4000				
£17 500				

C Simon has some money in a savings account. The rate of interest is 2%. He says: 'I will earn £12 interest over one year.'

How much does he have in his account?

D Julia is given £3500 and puts it in an account where it earns 3.2% annual interest.

Work out how much money the interest is worth on a monthly basis.

E Imagine you have £500 to put in a savings account for one year. Work out how much it will earn for different rates of interest. Show your results in a table or diagram.

11.5 Percentage increases and decreases

Learning objective

- To work out the result of a percentage change

Key words

| decrease | increase |
| reduce | reduction |

In a sale, prices are often described as being **reduced** by a certain percentage.

When the value of something **increases** or **decreases**, the change is often described using a percentage.

To work out the new value after a percentage change, you need to work out the percentage of the original value. You add it on for an increase. You subtract it for a decrease. A decrease is often called a **reduction**.

Example 8

This sign was shown in a shop window.

The original price of a dress is £119.95.

In the sale the price is reduced by 30%.

What is the sale price?

The reduction is 30% of £114.95.

30% of 114.95 = 0.3 × 114.95 = 34.49   The reduction is £34.49.

The sale price is £114.95 − £34.49 = £80.46.

SALE
NOW ON
Selected
prices
reduced
by 30%

Sometimes an increase can be more than 100%.

Example 9

The mass of a young puppy is 2.30 kg.

After a year its mass has increased by 180%.

What is its new mass?

180% of 2.30 = 1.8 × 2.30 = 4.14   The increase is 4.14 kg.

The new mass is 4.14 + 2.30 = 6.44 kg.

1 A bush is 1.85 m high. In a year the height increases by 15%.
    a Work out the increase in height.
    b Work out the new height.

2 A woman weighs herself.
    Her mass is only 43.4 kg and she is trying to increase it.
    She successfully increases her mass by 8%.
    a Work out the increase in her mass.
    b Work out her new mass.

(FS) 3 The price of a TV is £539.95.
    In a sale the price of the TV is reduced by 35%.
    Work out the new price.

(FS) 4 The price of a car is decreased by 3%.
    The original price was £14 895.
    Work out the new price.

5 A scientist has a population of fruit flies.
    To start with, there are about 150 flies.
    Two weeks later the number has increased by 250%.
    Work out how many flies the scientist has now.

(FS) 6 The price of a washing machine is £449. Work out the new price after an increase of:
    a 1.5%   **b** 2.6%   **c** 3.75%.

7 At some point in 2011 the population of the world reached 7 billion (7 000 000 000) and the rate of increase was 1.1% per year.
    Use those figures to estimate the population of the world exactly one year later.

(FS) 8 In a sale, prices are reduced.
    Work out the sale price of each of these items.

		Original price (£)	Reduction (%)
a	A jacket	85.50	20
b	A pair of trainers	63.99	35
c	A bag	89.50	70

(PS) 9 In 1901 the population of Peterborough was 30 872.
    In the next 70 years the population rose by 125.3%.
    Work out the population of Peterborough in 1971.

PS **10** This is an extract from a newspaper article in July 2011.

The Newtons' gas bill in 2011 was £715.37.

Their electricity bill was £518.52.

Work out the total increase in the Newtons' bills for gas and electricity in 2012.

GAS AND ELECTRICITY PRICES INCREASE

A fuel company has announced that domestic gas bills will rise by 17% and electricity bills will rise by 19%.

FS **11** A woman is earning £36 629 per year.

She is given a pay rise of 3.2%.

a Work out her new salary.

b She is paid $\frac{1}{12}$ of her salary each month. How much extra will she earn each month?

PS **12** The value of a house in 1990 was £120 000.

a In the next ten years the value increased by 50% of the value in 1990.
Work out the value in 2000.

b In the next ten years the value increases by 50% of the value in 2000.
Work out the value in 2010.

c If the same thing happens in the next ten years, estimate the value in 2020.

Financial skills: VAT

VAT is a tax that is added to goods and services in all countries in the European Union.

If you buy something or have work done by someone, VAT is probably included in your bill.

The rate of VAT can vary.

In January 2011 the rate of VAT in the UK increased from 17.5% to 20%.

A A garage charges £240 to repair a car.

The garage must add 20% VAT to the bill.

a Work out the VAT.

b Work out the total cost, including VAT.

B The cost of fitting a new kitchen, before adding VAT, is £8200.

If the rate of VAT is 17.5%, work out the total cost.

C When the rate of VAT went up from 17.5% to 20%, prices increased.

The cost of a camera before VAT is added is £354.

a If the rate of VAT is 17.5%, work out the VAT.

b Work out the increase in the price of the camera because of the increase in the VAT.

D The rate of VAT is only 5% on some items. This applies to domestic gas and electricity.

An electricity bill, before VAT is added, is £438.29.

a Work out the cost of VAT at 5%.

b Work out how much the bill would increase if the rate of VAT is increased to 20%.

Ready to progress?

Review questions

1   Write each of these fractions as a percentage.

a $\frac{5}{8}$ b $\frac{5}{7}$ c $\frac{5}{4}$ d $\frac{5}{3}$

2   Write each of these as a fraction, as simply as possible.

a   90%   b   95%   c   96%   d $\frac{1}{2}$%   e $2\frac{1}{2}$%

3   Write each percentage as a decimal.

a   90%   b   9%   c   0.9%   d   209%

4   Here are five numbers.

    1.6   159%   1.712 $\frac{7}{4}$ $\frac{13}{8}$

a   Write down the largest number.   b   Write down the smallest number.

5   Work out:

a $\frac{2}{3}$ of 60 cm   b $\frac{7}{9}$ of 60 cm   c $\frac{9}{5}$ of 60 cm   d $\frac{13}{3}$ of 60 cm

6   There are 1200 people in a concert hall. $\frac{2}{3}$ of them are teenagers. $\frac{3}{8}$ of the teenagers are boys.

(PS)

    How many teenage boys are there in the concert hall?

7   Work out:

a   40% of 750 kg   b   35% of 40 m   c   70% of 2.5 litres   d   125% of £26.00.

8   Work out these percentages of £47.53.

a   19%   b   1.8%   c   12.7%   d   109%

9   This pie chart shows the people in the crowd at a hockey game.

a   Estimate, to the nearest ten, the percentage of the crowd who are women.

b   There are 2372 people in the crowd.
    Estimate the number of women.

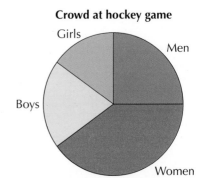

Crowd at hockey game

(PS) 10 This information was found on a website.

> Modern bronze is typically
> 88% copper and 12% tin

A small bronze statue has a mass of 157 g.

Work out the masses of copper and tin in the statue.

(FS) 11 Arthur is given £450. He spends 30% and donates 20% of what he has not spent to charity. How much does he have left?

12 This question is about the heights of buildings in London.

   a   Canary Wharf Tower is 243.8 m high. 30 St Mary Axe (known as the Gherkin) is 17.5% shorter than Canary Wharf Tower.
Work out its height.

   b   St Paul's Cathedral is 111.6 m high. The Shard is 177.7% taller.
Work out the height of The Shard.

(FS) 13 Max pays a bill of £429 with a credit card. There is a charge of 1.5% for doing this.

   a   What is the charge, in pounds?

   b   What is the total amount he has to pay?

   c   Suppose Max pays a bill of £N with his credit card.
Explain why the charge for using a credit card is £ $0.015N$.

14 Jan draws a rectangle. The side are 20 cm and 25 cm long.

   a   Work out the area of the rectangle.

   b   Lou draws another rectangle. Each side is 20% longer than Jan's. Work out:

      i   the lengths of the sides of Lou's rectangle   ii   the area of Lou's rectangle.

   c   Show that the area of Lou's rectangle is more than 40% greater than Jan's.

(FS) 15 Luis buys a very old car for £5250.

He restores it and sells it for 150% more than he paid for it.

Work out the price he sold it for.

(FS) 16 Josie's salary is £22 400 a year. She gets a pay rise of 2.7%.

Work out her salary after the increase.

17 On Friday a website gets 300 hits.

On Saturday the number of hits increases by 65% on Friday's hits.

   a   Work out the number of hits on Saturday.

   b   On Sunday the number of hits is 65% less than on Friday. Work out the number of hits on Sunday.

   c   Draw a bar chart to show the number of hits each day.

(PS) 18 Here are three sale notices.

Which one gives the greatest reduction?

Give a justification of your answer.

Financial skills

Income tax

Most people who earn money have to pay income tax.
You can earn a certain amount before you have to pay tax.
This is called your tax allowance.
You pay a percentage of anything over your tax allowance as income tax.
This percentage is called the tax rate.

Neil lives in Mathsland. He earns £25 000 in one year. His tax allowance is £8000.
The tax rate is 20%.

1 Copy and complete these sentences.

  a Neil pays tax on £25 000 − £8000 = £…

  b Neil's tax bill is 20% of … = £…

2 Work out the tax Neil will pay if he earns £35 000 in one year.

3 **a** Copy and complete this table for Neil. Fill in the answers from questions 1 and 2.

Income (£)	15 000	20 000	25 000	30 000	35 000	40 000
Tax (£)						

  b Use the table to draw a graph that shows how much tax Neil pays. Draw axes like this.

    Plot seven points and join them with a straight line that starts at 8000 on the 'income' axis.

4 Use your graph to find out how much tax Neil will pay if he earns £27 500 in one year.

5 One year Neil pays £5000 income tax. Use the graph to find out how much he earned.

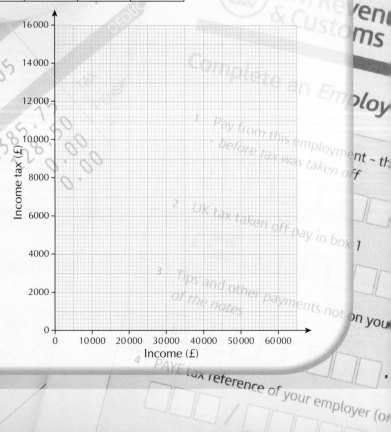

6 One year Neil earns £18 000. He says: 'If I earn twice as much, I shall pay twice as much income tax.'

Is this true? Give a reason for your answer.

For income over £40 000 Neil pays a higher rate of tax, at 40%.
For example, suppose Neil earns £50 000.

His tax allowance is £8000. ....He pays no tax on this.
He pays 20% of the next £32 000. ....That is £6400 tax.
He pays 40% of the final £10 000. ......That is £4000 tax.
If he earns £50 000 he will pay £6400 + £4000 = £10 400.

7 Copy and complete this table for income over £40 000.

Income (£)	45 000	50 000	55 000	60 000
Tax (£)		10 400		

8 Use your table to add more points to your graph. Join them up.

9 Neil pays £11 000 income tax. How much does he earn?

10 Neil says: 'If I earn £58 000, I shall pay over a quarter of my salary in income tax.'

Is this correct? Give a reason for your answer.

Investigate

Find out what the current tax allowances and rates are in the UK.

12

Probability

This chapter is going to show you:

- how to use the correct words about probability
- how to work with a probability scale
- how to work out theoretical probabilities in different situations
- how to use experimental probability to make predictions
- how to use sample space diagrams to work out the probability of a combined event
- how to compare experimental probabilities with theoretical probabilities.

You should already know:

- some basic ideas about chance and probability
- how to collect data from a simple experiment
- how to record data in a table or chart.

About this chapter

In October 2012 the mayor of New York had a tricky decision to make. Hurricane Sandy, a violent and destructive storm, was heading across the Atlantic towards the east coast of the USA. But where exactly would its main force hit? What effect would it have on New York?

As it approached, scientists calculated and recalculated the probability of it devastating the city. Eventually, with the storm just hours away, the mayor ordered the compulsory evacuation of 375 000 people and so saved many lives.

Assessing the probability of what might happen like this is vital for scientists trying to prevent natural events such as storms and earthquakes from turning into disasters for the people who might be in their way.

12.1 Probability scales

Learning objective

- To learn and use the correct words about probability

Key words

at random	biased
chance	equally likely
event	fair
outcome	probability
probability fraction	probability scale
random	

Look at the pictures. Which one has the greatest **chance** of happening where you live today?

Something that happens, such as the roll of a dice, is called an **event**.

The **outcomes** of the event are its possible results. For example, rolling a dice has six possible outcomes: a score of 1, 2, 3, 4, 5, or 6.

You can use **probability** to decide how likely it is that different outcomes will occur.

Here are some words you will hear, when talking about whether something may occur:

certain, very likely, likely, an even chance, unlikely, very unlikely, impossible.

The two complete opposites here are *impossible* and *certain*, with an even chance (evens) in the middle, and so these can be given in the order:

impossible, very unlikely, unlikely, evens, likely, very likely, certain.

These words can be shown on a **probability scale**.

Impossible Very   Unlikely Evens   Likely   Very   Certain
            unlikely   likely

There are lots of other words you can use when describing probability, such as:

50–50 chance, probable, uncertain, good chance, poor chance.

If you flip a coin or throw a dice the outcome is **random**, which means that the outcome cannot be predicted.

If you take a coloured ball from a bag without looking at it, you are 'taking the ball **at random**'.

Example 1

Match each of these outcomes to a position on a probability scale.

a It will snow in the winter in London.

b You will come to school in a helicopter.

c The next person to walk through the door will be male.

d When a normal dice is thrown the score will be 8.

a It usually snows in the winter, but not always, so this event is very likely.

b Unless you are very rich, this is very unlikely to happen.

c As the next person to come through the door will be either male or female this is an evens chance.

d An ordinary dice can only score from 1 to 6 so this is impossible.

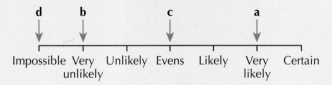

To measure probability as a value, rather than in words, you need to use a scale from 0 to 1. You can write probabilities as fractions or decimals, and sometimes as percentages, as in the weather forecasts.

The numerical probability scale looks like this.

0	0.1	0.2	0.3	0.4	0.5	0.6	0.7	0.8	0.9	1.0
Impossible					Evens					Certain

The probability of an outcome happening is written as:

$$P(\text{outcome}) = \frac{\text{number of ways the outcome can happen}}{\text{total number of all possible outcomes}}$$

The P stands for 'the probability of'. The outcome is shown inside the brackets.

Fair or biased

When you toss a **fair** coin, there are two possible outcomes: Heads (H) or Tails (T). When a coin, a dice or a spinner is described as fair it means each outcome is **equally likely** to occur. Sometimes, a coin may not be fair, so that one outcome is more likely than the rest. That means that it is **biased**.

Probability fraction

In the case of a fair coin, each outcome is equally likely. The number of ways each outcome can happen is 1, as the coin can only fall on one side. The total number of possible outcomes is 2, heads or tails. So:

$P(H) = \frac{1}{2}$ and $P(T) = \frac{1}{2}$

This is the **probability fraction** for the event.

The probability can also be written as $P(H) = 0.5$ or $P(H) = 50\%$.

When you throw a fair dice, there are six equally likely outcomes: 1, 2, 3, 4, 5, 6.

So, for example: $P(6) = \frac{1}{6}$ and $P(1 \text{ or } 2) = \frac{2}{6} = \frac{1}{3}$

Probability fractions can be simplified, but sometimes it is better to leave them unsimplified, especially if you are comparing probabilities.

Exercise 12A

1 Match each of these events to the probability outcomes.

Events

A You are older today than you were last year.

B You will score a prime number on an ordinary dice.

C You will walk on the moon tomorrow.

D Someone in your class will have a pet cat.

E Someone in your class will have a pet alligator.

F Someone in your class will have a birthday this month.

G Someone in your class will be an only child.

a certain	**b** impossible	**c** fifty-fifty chance	**d** very unlikely
e likely	**f** very likely	**g** unlikely	

2 Here are two sets of cards.

Set A: 1 2 3 2 2 3 2 4

Set B: 1 4 3 1 1 4 1 1

A card is picked at random. Copy and complete these sentences.

a Picking a … from set … is impossible.

b Picking a … from set … is likely.

c Picking a … from set … is unlikely.

d Picking a … from set … is very unlikely.

e Picking a … from set … is fifty-fifty.

3 Here are two sets of shapes.

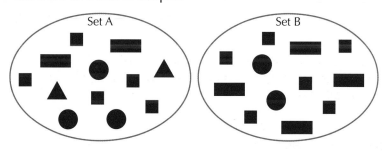

Set A   Set B

a A shape is chosen at random from set A. What is the probability that it is:

    i a triangle   **ii** a circle   **iii** a square   **iv** a rectangle   **v** a hexagon?

b A shape is chose at random from set B. What is the probability that it is:

    i a triangle   **ii** a circle   **iii** a square   **iv** a rectangle   **v** blue?

Give your answers as fractions in their simplest form.

4 Cards numbered from 1 to 10 are placed in a box. A card is drawn at random from the box. Find the probability that the card drawn is:

a 8   **b** an odd number   **c** a multiple of 4

d A factor of 10   **e** a prime number   **f** a square number.

Give your answers as decimals.

5 A bag contains five green discs, three white discs and two orange discs. Linda takes out a disc at random. Find the probability that she takes out:

a a green or white disc   **b** a black disc

c a green or red disc   **d** a disc of a colour that is on the Irish flag.

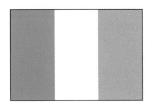

6 Sandhu rolls an ordinary fair dice. Find:

a P(2)   **b** P(square number)   **c** P(factor of 6)

d P(prime number)   **e** P(1 or 6)   **f** P(factor of 60).

7 Mr Evans has a box of 30 calculators, but six of them do not work very well.

He takes one out of the box at random. What is the probability that this calculator does not work very well?

Write your fraction as simply as possible.

8 At the start of a fair, there are 500 tickets inside the tombola drum. There are 75 winning tickets available.

What is the probability that the first ticket taken out of the drum is a winning ticket?

Write your fraction as simply as possible.

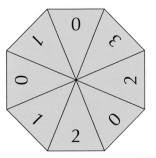

(MR) 9 Brian rolls a dice 600 times. He scores a six 180 times.

Do you think that the dice is biased? Explain your answer.

(MR) 10 a Syed is using a fair, eight-sided spinner in a game. Find the probability that he scores:

     i 0   **ii** 1   **iii** 2   **iv** 3.

b Syed now designs his own fair, eight-sided spinner. He writes a numbers from 1 to 4 on each section. Copy a blank spinner from part **a** and fill in the numbers so that:

$P(1) = \frac{1}{8}$, $P(2) = \frac{1}{4}$, $P(3) = \frac{1}{2}$.

Activity: Play your cards right

Work in pairs.

You will need a set of cards numbered 1 to 10 for this experiment.

A One of you shuffles the pack, deals out five cards, face down, and then puts the rest to one side, face down. You will use these when you swap roles with your partner.

   1 Turn over the first card.

   2 Ask your partner to write down the probability that the number on the second card will be higher or lower than the one on this card. They should give the probability for their prediction.

   3 Turn over the second card.

   4 If your partner was correct, then ask them to make a prediction and work out the probabilities for the number on the third card being lower or higher than the number on the second card. If they are wrong their game is ended.

   5 Repeat with the remaining cards.

B Play the game 10 times and record how many times your partner managed to predict correctly, up to the fifth card.

C Swap over and repeat the activity.

D If the whole class have done the activity, discuss any strategies you used to predict the numbers.

   a Are there some numbers that make it certain your prediction will be correct?

   b Are there some numbers that make it likely your prediction will be correct?

   c Are there some numbers that make it hard to predict what the next one will be?

   d Does it get easier to make correct predictions after you see the first couple of cards?

12.2 Combined events

Learning objective

- To use sample space diagrams to work out the probability of a combined event

Key words

combined event

list of outcomes

sample space diagram

In a **combined event**, two separate events are linked in some way. For example, a coin and an ordinary dice can be thrown together, so there are 12 outcomes, such as Heads and a 6, Tails and a 4.

You can use a **list of outcomes** or a **sample space diagram** to show the equally likely outcomes for two combined events.

Example 2

Tom has three pairs of jeans, which are white, blue and black.

He also has four T-shirts, which are white, red, blue and black.

Today he is wearing the blue jeans and the blue T-shirt.

a List all the possible combinations of jeans and T-shirts that Tom could wear.

b If Tom chooses his jeans and T-shirt randomly, what is the probability that he wears jeans and a T-shirt that are the same colour?

    a Write the combinations with jeans first.

      (white, white), (white, red), (white, blue), (white, black), (blue, white), (blue, red), (blue, blue), (blue, black), (black, white), (black, red), (black, blue), (black, black)

    b There are 12 possible combinations.

      Three of these give the same colour jeans and T-shirt, so the probability of this is:

      $\frac{3}{12} = \frac{1}{4}$

Example 3

This sample space diagram shows all the possible outcomes for the total score when two fair dice are thrown.

Work out: **a** P(double 6) **b** P(score of 8) **c** P(score greater than 10).

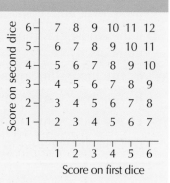

    The sample space diagram shows that there are 36 equally likely outcomes.

    a P(double 6) = P(score of 12) = $\frac{1}{36}$

    b P(score of 8) = $\frac{5}{36}$

    c P(score greater than 10) = P(score of 11 or 12) = $\frac{3}{36} = \frac{1}{12}$

Exercise 12B

1 The sample space diagram shows the outcomes when two fair coins are tossed.

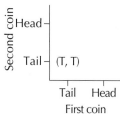

    a Copy and complete the diagram.

    b Use your completed sample space diagram to find:

      i P(H, H)   **ii** P(T, T)   **iii** P(1 Head and 1 Tail)   **iv** P(not getting 2 Heads)

2 Marilyn says that the result of a football match can be win, lose or draw. She then says that, since there are three outcomes…

... the probability that a team wins a match is $\frac{1}{3}$.

Is she correct? Explain your answer.

 3 I need to score 6 or more to win a game. I can choose to roll one or two dice. Which option should I choose? Use the sample space diagram in Example 3 to help you.

Explain your answer.

4 You have five rods, A, B, C, D and E, with lengths of 10 cm, 20 cm, 30 cm, 40 cm and 50 cm.

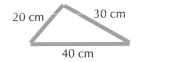

You pick up three rods at random and, if it is possible, make them into a triangle.

For example if you pick B, C and D you can make make a triangle. If you pick A, B and E you cannot make a triangle.

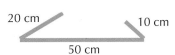

There are 10 different combinations of three rods that can be made from the five rods, for example, ABC, ABE, DCE.

a Write down the 10 different combinations.

b Is it more likely that you will pick three that will make a triangle, or you will pick three rods that will not make a triangle? Explain your answer fully.

 5 Currently British coins with values under a pound (pence, p) are worth 1p, 2p, 5p, 10p, 20p and 50p.

Currently American coins with values under a dollar (cents, c) are worth 1c, 5c, 10c, 25c and 50c.

Dave has two British coins in his pocket.

Barack has two American coins in his pocket.

a List all the possible combinations of coins that Dave could have in his pocket. There are 21 possible totals.

b **i** List all the possible combinations that Barack could have in his pocket.

   ii How many possible combinations are there?

c A Snacker chocolate bar costs 45p in Britain and 45c in America.

Who is more likely to have the right coins to buy a Snacker bar?

Explain your answer.

6 Ryan spins the two fair, five-sided spinners shown here. He records the product of their scores.

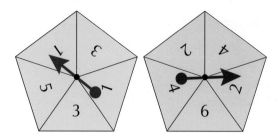

Hint The product is the result of multiplying the numbers together.

a Draw a sample space diagram to show all the possible outcomes.

b How many equally likely outcomes are there?

c Work our each of these probabilities.

   i P(score of 18)   **ii** P(score of 4)

   iii P(score that is a multiple of 3)   **iv** P(odd score)

7 **a** Draw a sample space to show all the possible outcomes for the total score from two normal fair dice.

b Find the probability of rolling a total that is:

   i less than 6   **ii** equal to 6   **iii** greater than 6.

c **i** Add together the three probabilities in part **b**.

   ii Comment on your result.

d **i** What is the probability of rolling a total of 7?

   ii What is the probability of rolling a total that is **not** 7?

 8 You have two dice:

Dice A showing 0, 1, 1, 2, 2, 3

Dice B showing 0, 2, 3, 3, 4, 5

a Draw a sample space to show all the possible outcomes for the **product** of the two dice.

b Find the probability of rolling a product of 6.

c Find the probability of rolling:

   i an even product   **ii** an odd product.

d Comment on the sum of the probabilities in part **c**.

e **i** What is the probability of rolling a product of 4?

   ii What is the probability of rolling a product that is **not** 4?

9 Tegan spins two fair, four-sided spinners together. Each spinner is numbered 1, 2, 3, 4.

If the scores on the two spinners are both odd, Tegan adds the scores together.

If one or both of the scores is even, Tegan works out their product.

a Copy and complete the sample space diagram to show Tegan's outcomes.

		Score on spinner 1			
		1	2	3	4
	1	2	2	4	
Score on spinner 2	2				
	3				
	4				

b What is the probability that the combined score will be a square number?

c What is the probability that the combined score will be a prime number?

d Explain why the probability of the combined score being odd is zero.

10 A bag contain 30 balls, which are either red or white. There are twice as many white balls as red.

a Yashmin says: 'The probability of taking a red ball from the bag at random must be $\frac{1}{2}$.'

Explain why she is wrong.

b More red balls are added to the bag so that, when a ball is taken from the bag at random, P(red) = $\frac{1}{2}$. How many red balls were added? Show your working.

Activity: Scores on a spinner

A Copy and complete the table to show for the number of times you would expect a fair, six-sided spinner to land on each number, in 60 spins.

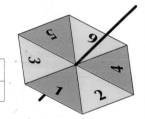

Score	1	2	3	4	5	6
Frequency	10					

B a Make a six-sided spinner from card and a cocktail stick.

b Spin it 60 times and record the scores in a table like the one above.

C a Now weight it by sticking a small piece of sticky gum below one of the numbers on the card.

This will make the spinner unfair or biased.

b Spin the spinner 60 times and record the scores in a table like the one above.

D Compare your three sets of results:

- what you would expect from a fair, six-sided spinner
- your scores with your unbiased spinner
- your scores with the biased spinner.

12.3 Experimental probability

Learning objectives

- To understand experimental probability
- To understand the difference between theoretical probability and experimental probability

Key words

experimental probability

theoretical probability

trial

You calculated the probabilities in the previous section by using equally likely outcomes. A probability worked out this way is a **theoretical probability**.

Sometimes, you can only find a probability by carrying out a series of experiments and recording the results in a frequency table. Then you use these results to estimate the probability of an outcome. A probability found in this way is an **experimental probability**.

To find an experimental probability, you must repeat the experiment a number of times. Each separate experiment carried out is a **trial**.

$$\text{Experimental probability of an outcome} = \frac{\text{number of times the outcome occurs}}{\text{total number of trials}}$$

It is important to remember that, when you repeat an experiment, the experimental probability will be slightly different each time. The experimental probability of an event is an estimate for the theoretical probability.

As the number of trials increases, the value of the experimental probability gets closer to the theoretical probability.

Example 4

When Gianni spun 20 coins, 12 of them landed on heads. What is the experimental probability of a coin landing on heads?

The experimental probability of a coin landing on heads in this case is $\frac{12}{20} = \frac{3}{5}$.

Example 5

Hassan's mother wanted to know the probability of the school bus coming late. Over three weeks, she counted that the bus came late five times. What is her estimate of the probability that the school bus will come late?

The bus was late 5 out of 15 days, so her experimental probability is $\frac{5}{15} = \frac{1}{3}$.

Example 6

A company manufactures items for computers. The number of faulty items is recorded in this table.

Number of items produced	Number of faulty items	Experimental probability
100	8	0.08
200	20	
500	45	
1000	82	

a Copy and complete the table.

b Which is the best estimate of the probability of an item being faulty? Explain your answer.

a

Number of items produced	Number of faulty items	Experimental probability
100	8	0.08
200	20	0.1
500	45	0.09
1000	82	0.082

b The best estimate is the last result (0.082), as this is based on more results.

Exercise 12C

1 Flip a coin 20 times and record your results in a frequency table.

   a Use your results to find your experimental probability of getting a Head.

   b What is the theoretical probability of getting a Head?

   c How many Heads would you expect to get after tossing the coin 100 times?

2 Roll a dice 30 times and record your results in a frequency table.

   a Find your experimental probability of scoring 6, writing your answer as:

      i a fraction   **ii** a decimal.

   b The theoretical probability of getting a 6 is $\frac{1}{6}$ or 0.17. How close is your experimental probability to the theoretical probability?

   c Explain how you could improve the accuracy of your experimental probability.

3 Drop a drawing pin 20 times. Record your results in a frequency table.

	Tally	Frequency
Point up		
Point down		

   a Find your experimental probability of the drawing pin landing point up, writing your answer as:

      i a fraction   **ii** a decimal.

   b Luke dropped a box of 100 drawing pins and they all fell out onto the floor. How many of them should Luke expect to land point up?

4 Brian says: 'When I drop a piece of toast, it always lands butter-side down.'

Test Brian's statement by dropping a playing card 50 times and recording the number of times the card lands face down.

   a What is the experimental probability that the card lands face down?

   b Do you think this is a good way to test Brian's statement? Explain your answer.

5 Working in pairs, flip three coins together 50 times. Record, in a frequency table, the number of Heads you get for each trial.

  a What is the experimental probability of getting:

   i three Heads　　**ii** two Heads　　**iii** one Head　　**iv** no Heads?

  b List all the equally likely outcomes for flipping three coins.
  What is the theoretical probability of scoring:

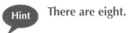 **Hint** There are eight.

   i three Heads　　**ii** two Heads　　**iii** one Head　　**iv** no Heads?

 6 Suzie suspects a dice is biased. She throws it 120 times. The table shows her results.

Score	1	2	3	4	5	6
Frequency	19	6	22	21	36	16
Experimental probability						

  a How does this suggest that Suzie's suspicions are correct? Explain your answer.

  b Fill in the experimental probability row. Give your answers as decimals to two decimal places (2 dp).

  c Which number is the dice biased towards? Explain how you can tell.

 7 A bag contains 10 coloured balls. Zeena takes a ball from the bag at random. She puts the ball back and repeats this nine times.

  The colours she took from the bag were:

   red, red, blue, red, blue, red, blue, blue, red, red.

  Which of these statements are *definitely true* (T), which are *definitely false* (F) and which *may be true* (M)?

  A: There are only red and blue balls in the bag.

  B: There are more red balls than blue balls.

  C: The probability of a taking a red ball at random is from 0.1 to 0.9 inclusive.

  D: The probability of a taking a blue ball at random is 0.05.

 8 **a** Jake threw a biased dice. He recorded the number of fours.
   After 10 throws the experimental probability P(4) was 0.3.
   Explain how you know that he threw three fours in the first ten throws.

  b After 20 throws he had thrown 7 fours in total. Work out the experimental probability P(4) after 20 throws.

  c After 30 throws the experimental probability P(4) was 0.4.
   How many times did Jake throw a four between the 20th and 30th throws?

  d After 40 throws, Jake calculated that the experimental probability P(4) was 0.25.
   Explain why Jake must have made a mistake.

Investigation: Toss the stick

You will need a matchstick or cocktail stick and a piece of plain, A4 paper.

A Mark the paper with parallel lines that are the width of the matchstick or cocktail stick apart.

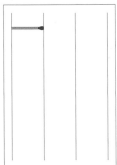

B a Now drop the match or cocktail stick from a height of about 1 metre, down onto the paper.

Record if the match or cocktail stick touches a line or does not touch a line.

No   Yes

b Do this at least 100 times and work out the experimental probabilities P(does not touch) and P(does touch).

Work out each of your answers as a decimal.

C Now work out 2 ÷ P(does touch).

It has been shown mathematically that this value should equal π (3.142).

How far away was your answer?

D If possible combine all of the results of the class and see if a larger number of experiments gives a value closer to π.

Ready to progress?

I can use a probability scale in words.
I can use a probability scale marked from 0 to 1.

I can use equally likely outcomes to calculate probabilities.
I can calculate probability from experimental data.
I understand the differences between theoretical and experimental probability.

I can calculate probabilities from sample spaces and lists.
I can compare experimental probabilities with theoretical probabilities.

Review questions

1 **a** Eve puts five white counters and three blue counters in a bag.

   She is going to take out one counter, without looking.

   What is the probability that the counter will be blue?

   b Eve puts the counter back in the bag and then puts more blue counters in the bag.

   She is going to take one counter without looking.

   The probability that the counter will be blue is now $\frac{2}{3}$.

   How many more blue counters did Eve put in the bag?

2 Jo and Lucy are playing a game of 'Hit the rat'. They each shade in two adjacent squares on a 4 by 4 grid without showing each other. This is where the 'rat' is. Then they take it in turns to guess the squares that the other person's rat is in. The first to get both squares is the winner.

   a This is Lucy's board.

   What is the probability that Jo guesses one of the correct squares with his first guess?

   b Jo's first guess is B2 and he hits one of Lucy's rat squares.

   Explain why the probability that Jo gets the other square on his next go should be $\frac{1}{4}$.

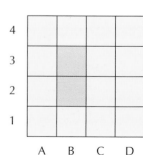

MR **3** The table shows the results, when Helen rolled a dice 100 times.

Number on dice	1	2	3	4	5	6
Frequency	13	15	18	16	15	23

    a Does this suggest that Helen's dice is biased? Explain your answer.

    b What is the theoretical probability of rolling each of these numbers?

    c If you rolled a fair dice 600 times, write down the kind of experimental probabilities you would expect to get for each number.

PS **4** A box of crisps contains three different flavours.

    A quarter of the packets are roast chicken flavour.

    The probability of picking a salt'n'vinegar flavour is 35%.

    What is the probability of picking a cheese'n'onion flavour?

MR **5** Dan wrote the numbers 1, 2, 3, 4 and 5 on a set of five cards. He shuffled them, placed them face down in a row and then turned over the first card.

    Dan challenged Beth to predict whether the number on the next card would be higher or lower.

    She guessed correctly for the following two cards.

    Explain why Beth should definitely guess the next two cards correctly.

MR **6** Meggie puts 10 coloured balls in a bag. The balls are either red (R) or blue (B). Meggie takes out a ball at random, notes its colour and then puts it back in the bag. After she has done this 100 times the experimental probabilities are P(R) = 0.38 and P(B) = 0.62.

    A blue ball is then taken from the bag and discarded so there are only nine balls left in the bag.

    Meggie takes out a ball at random, notes its colour and then puts it back in the bag. Again, she does this 100 times.

    Which of the following is the most likely set of experimental probabilities now?

    A: P(R) = 0.43, P(B) = 0.57   B: P(R) = 0.5, P(B) = 0.5   C: P(R) = 0.57, P(B) = 0.43

    Explain your choice.

Financial skills
School Easter Fayre

Westfield Community School is having an Easter Fayre to raise money.

The attractions included a tombola stall and a roller dice game.

Tombola

Parents have donated 100 prizes.

Each prize has a ticket attached to it.

These tickets are numbered 5, 10, 15 20, ... up to 500.

There are 500 tickets to sell, numbered from 1 to 500, rolled up in straws.

A straw costs 50p. If the number in the straw ends in 0 or 5 then it wins the prize with that number on it.

Use the information about the tombola stall to answer these questions.

1 If all the straws are sold, how much money will the tombola stall make?

2 Mr Jackson is the first to buy a straw. What is the probability of him winning a prize?

3 Mrs Wilson buys six straws. Her daughter Jackie buys two.

   Who has the greater chance of winning a prize?

4 Later in the day, 400 straws have been sold and 75 prizes have been won.

   Is the chance of winning a prize now greater or smaller than it was at the start of the day?

   Explain your answer.

5 15 minutes before the fayre ends, the tombola has taken £247.50 and the probability of winning a prize from the remaining tickets is 0.6.

   How many prizes are left? Show your working.

Roller dice

Class 9K are charging £1 for players to roll two dice.

If the player scores no sixes they win a creme egg.

If the player scores one six they get their money back.

If the player scores two sixes they win £2.50.

They expect to have 240 players.

Use the information about the roller dice stall to answer these questions.

Hint Draw a sample space diagram for the outcomes of throwing two dice.

6 During the afternoon, 252 people play the game.

  a Explain why you would expect seven of them to throw two sixes.

  b How many would you expect to throw:

  i one six   ii no sixes?

  c How much money will the class take without allowing for prizes?

  d Show that if creme eggs cost 40p the class will expect to make £94.50.

7 The headteacher plays the game.
  She says: 'It must be one of three prizes, so the probability that I win £2.50 is $\frac{1}{3}$.'

  Is she correct? Explain your answer.

13

Symmetry

This chapter is going to show you:
- how to recognise shapes that have reflective symmetry
- how to use line symmetry
- how to recognise and use rotational symmetry
- how to reflect shapes in a mirror line
- how to use coordinates to reflect shapes in all four quadrants
- how to rotate a shape
- how to tessellate a shape.

You should already know:
- how to recognise symmetrical shapes
- how to plot coordinates.

About this chapter

If you drew one side of an animal, such as a pet dog or cat, could someone draw the other side without looking at the animal? They probably could, because, for most animals, one side of the body reflects the other, like a mirror image. Their legs are the same length, their feet and ears are the same size. Their bodies are symmetrical. There are many other examples of symmetry all around us – in leaves, flowers, feathers. Symmetry is essential in the living world.

13.1 Line symmetry and rotational symmetry

Learning objectives

- To recognise shapes that have reflective symmetry and draw their lines of symmetry
- To recognise shapes that have rotational symmetry and find the order of rotational symmetry

Line symmetry

A 2D shape has a **line of symmetry** when one half of the shape fits exactly over the other half when the shape is folded along that line.

A mirror or tracing paper can be used to check whether a shape has a line of symmetry.

Some shapes have no lines of symmetry.

A line of symmetry is also called a **mirror line** because the shapes on each side reflect each other. This is called **reflective symmetry**.

Example 1

Describe the symmetry of this shape.

This T-shape has one line of symmetry, as shown.

Put a mirror on the line of symmetry and check that the image in the mirror is half the T-shape.

Trace the T-shape and fold the tracing along the line of symmetry to check that both halves of the shape fit exactly over each other.

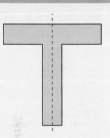

Example 2

Describe the symmetry of this shape.

This cross has four lines of symmetry, as shown.

Check that each line drawn here is a line of symmetry.

Use either a mirror or tracing paper.

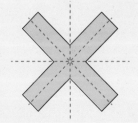

Example 3

Describe the symmetry of this shape.

This L-shape has no lines of symmetry.

You can check by using a mirror or tracing paper.

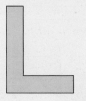

A 2D shape has **rotational symmetry** when it can be rotated about a point to look exactly the same in a new position.

The **order of rotational symmetry** is the number of different positions in which the shape looks the same when it is rotated about the point through one complete turn (360°).

A shape has no rotational symmetry when it has to be rotated through one complete turn to look exactly the same. So it is said to have rotational symmetry of order 1.

To find the order of rotational symmetry of a shape, use tracing paper.

- First, trace the shape.
- Then rotate the tracing paper until the tracing again fits exactly over the shape.
- Count the number of times that the tracing fits exactly over the shape until you return to the starting position.
- The number of times that the tracing fits is the order of rotational symmetry.

Example 4

Describe the symmetry of this shape.

This shape has rotational symmetry of order 3.

Example 5

Describe the symmetry of this shape.

This shape has rotational symmetry of order 4.

Example 6

Describe the symmetry of this shape.

This shape has no rotational symmetry.

Therefore, it has rotational symmetry of order 1 as its fits on top of itself only once.

Exercise 13A

1 Write down the number of lines of symmetry for each of these road signs.

a

b

c

d

e

f

2 **a** Write down the number of lines of symmetry for each shape.

i

ii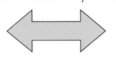

iii

iv

v

vi

vii

viii

b Write down the order of rotational symmetry for each shape.

3 The points A(4, 1), B(2, 3) and C(6, 3) are shown on the grid.

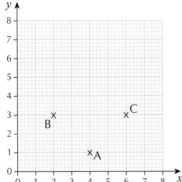

Copy the grid onto squared paper and plot the points A, B and C.

a Plot a point D so that the four points have no lines of symmetry.

b Plot a point E so that the four points have exactly one line of symmetry.

c Plot a point F so that the four points have exactly four lines of symmetry.

PS **4** Here are four identical squares.

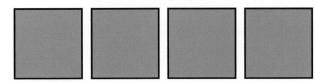

Copy the four squares and put them together to make a shape that has:

a no lines of symmetry **b** exactly one line of symmetry
c exactly two lines of symmetry **d** exactly four lines of symmetry.

 5 Here is a sequence of symmetrical shapes.

Use line symmetry to help draw the next two shapes in the pattern.
Explain your rule.

6 Write down the order of rotational symmetry for each of the shapes below.

a **b** **c**

d **e**

7 Copy and complete the table for each of these regular polygons.

a **b** **c** **d**  **e**

Shape	Number of lines of symmetry	Order of rotational symmetry
a Equilateral triangle		
b Square		
c Regular pentagon		
d Regular hexagon		
e Regular octagon		

What do you notice?

8 These patterns are from Islamic designs.

Write down the order of rotational symmetry for each design.

a  **b**  **c**

(PS) 9 Here are four identical rhombi.

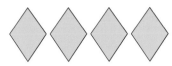

Put the four rhombi together to make a shape that has rotational symmetry of:

a order 1   **b** order 2   **c** order 4.

Investigation: Symmetrical triangles

Here are two identical equilateral triangles.

They can be placed together to make different symmetrical shapes.

This symmetrical shape has two lines of symmetry and rotational symmetry of order 2.

This symmetrical shape has two lines of symmetry and rotational symmetry of order 2.

A Investigate how many different symmetrical arrangements there are for three equilateral triangles.

Use isometric paper to draw your shapes.

B What arrangements can you make with four equilateral triangles?

13.2 Reflections

Learning objectives

- To understand how to reflect a shape
- To use coordinates to reflect shapes in all four quadrants

Key words

image	object
reflect	reflection

The diagram shows an L-shape **reflected** in a mirror.

You can use a mirror line to draw the **reflection**, like this.

The **object** is reflected in the mirror line to give the **image**.

The mirror line becomes a line of symmetry. So, if the paper is folded along the mirror line, the object will fit exactly over the image. The image is the same distance from the mirror line as the object is.

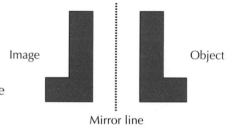

Image   Object

Mirror line

Example 7

Describe the reflection in this diagram.

Triangle A'B'C' is the reflection of triangle ABC in the mirror line.

Notice that the line joining A to A' is at right angles to the mirror line and the three points on the object and image are at the same distance from the mirror line.

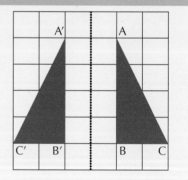

Example 8

Reflect this rectangle in the diagonal mirror line shown.

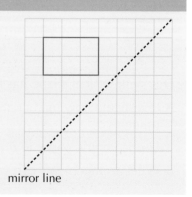

mirror line

Use tracing paper or a mirror to check the reflection.

Notice that the points joining the four vertices are perpendicular to the mirror line. Turning the diagram until the mirror line is vertical will make this easier to see.

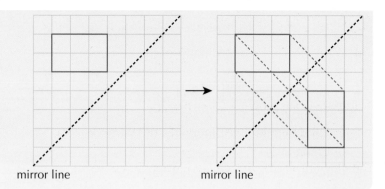

mirror line   mirror line

Example 9

Reflect the rectangle in the two mirror lines.

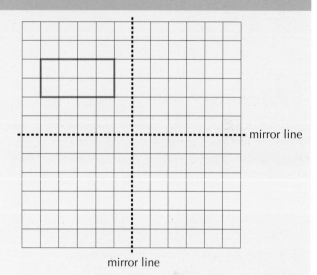

mirror line

mirror line

First reflect the rectangle in the vertical mirror line and then reflect both rectangles in the horizontal mirror line.

mirror line

mirror line

Example 10

a Write down the coordinates of the vertices of triangle P.

b Reflect P in the x-axis. Label the new triangle Q.

c Write down the coordinates of the vertices of triangle Q.

d Reflect Q in the y-axis. Label the new triangle R.

e Write down the coordinates of the vertices of triangle R.

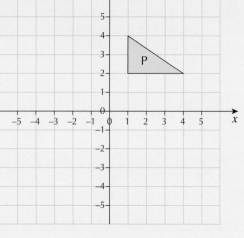

a (1, 4), (4, 2), (1, 2)

b and d See the diagram.

c (1, −4), (4, −2), (1, −2)

e (−1, −4), (−4, −2), (−1, −2)

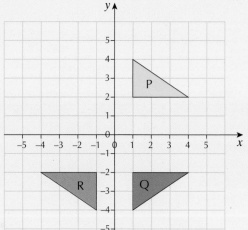

Exercise 13B

1 Copy each diagram onto squared paper and draw its reflection in the dotted mirror line.

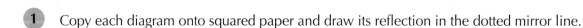

a

b

c

d

2 Copy each shape onto squared paper and draw its reflection in the dotted mirror line.

a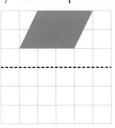

b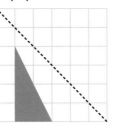

c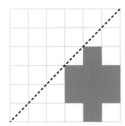

d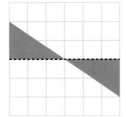

3 **a** Copy the coordinate grid onto squared paper and draw the shape and the mirror line.
Reflect the shape in the mirror line.
Label the vertices P', Q', R' and S'.

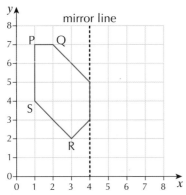

b Copy and complete the table.

Object		Image	
Vertex	Coordinates	Vertex	Coordinates
P		P'	
Q		Q'	
R		R'	
S		S'	

4 Copy each shape onto squared paper.
Reflect the shape in mirror line 1. Then reflect the shapes in mirror line 2.

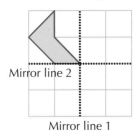

Mirror line 2   Mirror line 2   Mirror line 2

Mirror line 1   Mirror line 1   Mirror line 1

5 **a** Copy the diagram onto squared paper and reflect the shape in the series of parallel mirror lines.

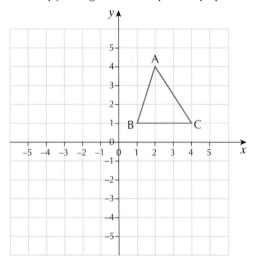

    b Use a series of parallel mirrors to make up your own patterns.

6 **a** Copy the grid onto squared paper and draw the triangle ABC.

Write down the coordinates of A, B and C.

  b Reflect the triangle in the *x*-axis. Label the vertices of the image A′, B′ and C′. What are the coordinates of A′, B′ and C′?

  c Reflect triangle A′B′C′ in the *y*-axis. Label the vertices of this image A″, B″ and C″. What are the coordinates of A″, B″ and C″?

  d Reflect triangle A″B″C″ in the *x*-axis. Label the vertices A‴, B‴ and C‴. What are the coordinates of A‴B‴C‴?

  e Describe the reflection that moves triangle A‴B‴C‴ onto triangle ABC.

Activity: Reflecting shapes

Use ICT software to reflect shapes in mirror lines.

13.3 Rotations

Learning objective

• To understand how to rotate a shape

You have seen how a 2D shape can be **rotated**. To describe the **rotation** of a 2D shape fully, you need to know three facts:

• **centre of rotation** – the point about which the shape rotates

• **angle of rotation** – usually 90° (a quarter-turn), 180° (a half-turn) or 270° (a three-quarter turn)

• **direction of rotation** – clockwise (to the right) or anticlockwise (to the left).

When you rotate a shape, it is a good idea to use tracing paper.

As with reflections, you call the original shape the object and the rotated shape the image.

Example 11

Describe this rotation.

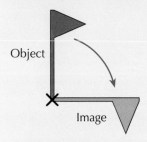

The flag has been rotated through 90° clockwise about the point X.

Notice that this rotation gives the same image as rotating the flag through 270° anticlockwise about the point X.

Example 12

Describe this rotation.

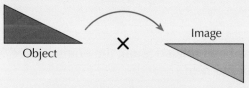

The right-angled triangle has been rotated through 180° clockwise about the point X.

Notice that the triangle (or any other shape) can be rotated either clockwise or anticlockwise when the angle of rotation is 180°.

Example 13

Describe this rotation.

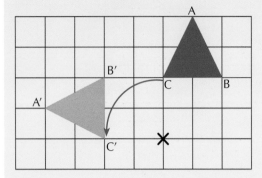

Triangle ABC has been rotated through 90° anticlockwise about the point X onto triangle A'B'C'.

Example 14

Describe this rotation.

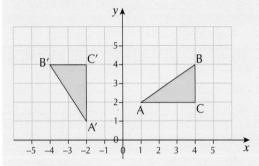

Triangle A'B'C' is the image of triangle ABC after a rotation of 90° anticlockwise about the origin O.

The coordinates of the vertices of the object are A(1, 2), B(4, 4) and C(4, 2) and the coordinates of the vertices of the image are A'(−2, 1), B'(−4, 4) and C'(−2, 4).

Exercise 13C

1 Copy each flag and draw its image after it has been rotated about the point marked X, through the angle indicated. Use tracing paper to help.

a   b    c   d

90° anticlockwise   180° clockwise   90° clockwise   270° anticlockwise

2 Copy each shape onto a square grid. Draw the image after the shape has been rotated about the point marked X, through the angle indicated. Use tracing paper to help.

a

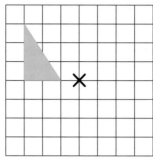

180° clockwise

b

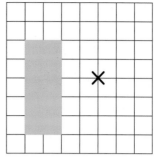

90° anticlockwise

c

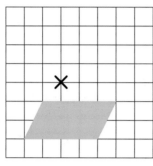

180° anticlockwise

d

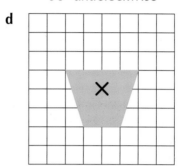

90° clockwise

3 Copy the triangle ABC onto a coordinate grid, with axes for *x* and *y x* numbered from −5 to 5.

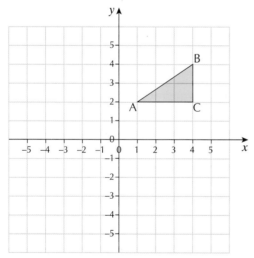

a Write down the coordinates of ABC.

b Rotate triangle ABC through 90° anticlockwise about the origin O.

c Write down the coordinates of the vertices of image A'B'C'.

d Rotate triangle A'B'C' through 90° anticlockwise about the origin O.

e Write down the coordinates of the vertices of image A"B"C".

f Describe the rotation that moves triangle A"B"C" back onto the original triangle ABC.

4 Draw a coordinate grid, with both axes numbered from −5 to 5. Copy each of these shapes onto your grid.

    a Draw the image of each one after it has been rotated about the origin (0, 0), through the angle and direction indicated.

    b Write down the coordinates of the vertices of each object and its image.

i 180° clockwise

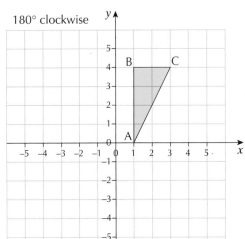

ii 90° anticlockwise

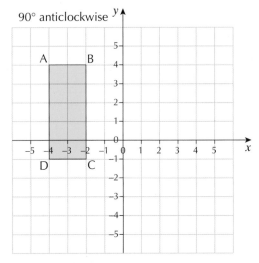

iii 180° anticlockwise

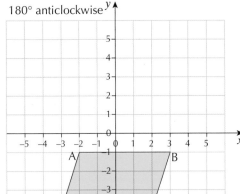

iv 180° anticlockwise

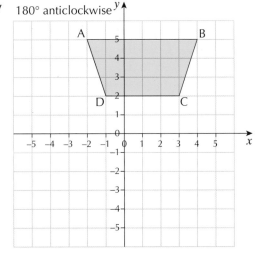

5 Draw a coordinate grid, with both axes numbered from 0 to 6.

Copy this rectangle onto your grid.

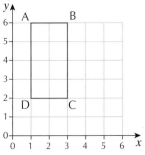

    a Rotate the rectangle ABCD through 90° clockwise about the point (1, 2), to give the image A′B′C′D′.

    b Write down the coordinates of A′, B′, C′ and D′.

    c Which coordinate point remains fixed throughout the rotation?

    d What rotation will move the rectangle A′B′C′D′ onto the rectangle ABCD?

Draw a coordinate grid, with both axes numbered from −5 to 5.

Copy this trapezium onto your grid.

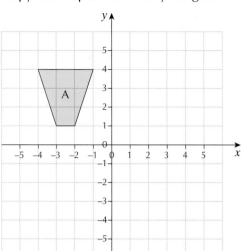

a Rotate trapezium A through 270° clockwise about the point (2, 1). Label the image B.

b Write down the coordinates of the vertices of trapezium B.

c Write down two different rotations that will move trapezium B onto trapezium A.

Challenge: Rotating rectangles

The grid shows three rectangles.

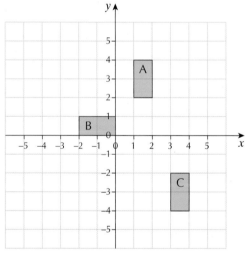

> **Hint** The centre of rotation is different for each one.

A Describe the rotation that moves A onto B.

B Describe the rotation that moves A onto C.

C Describe the rotation that moves C onto B.

13.4 Tessellations

Learning objective

- To understand how to tessellate shapes

Key word

tessellation

A **tessellation** is a pattern made by fitting together identical shapes without leaving any gaps.

When drawing a tessellation, use a square or a triangular grid, as in the examples below.

To show a tessellation, you should generally draw up to about ten repeating shapes.

Example 15

These are two different shapes that each makes a tessellation on a square grid.

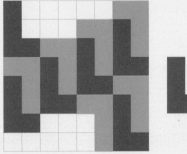 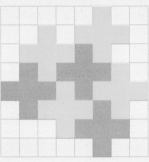

Example 16

This shape tessellates on a triangular grid.

Example 17

Circles do not tessellate.

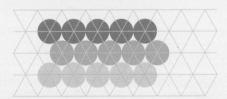

However you try to fit circles together, there will always be gaps.

Exercise 13D

1 Make a tessellation for each shape. Use a square grid.

a **b**  **c** **d**

2 Make a tessellation for each shape. Use a triangular grid.

a **b** **c** **d**

(PS) **3** Investigate which of these regular polygons will tessellate.

a **b** **c** **d**

e

4 Do all quadrilaterals tessellate?

This is an example of an irregular quadrilateral.

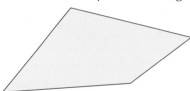

Make ten identical irregular quadrilateral tiles cut from card.

Then use your tiles to see if they tessellate.

5 Here is a tessellation that uses curves.

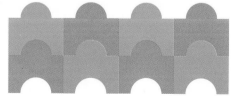

Design a different tessellation that uses curves.

Activity: Design a tessellation

Work with a partner or in a group.

A Design a tessellation of your own.

B Make an attractive poster to show all your different tessellations.

Ready to progress?

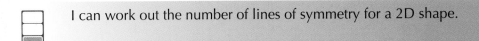

I can work out the number of lines of symmetry for a 2D shape.

I can work out the order of rotational symmetry of a 2D shape.
I can reflect a 2D shape in horizontal, vertical and diagonal mirror lines.
I can tessellate 2D shapes.

I can use coordinates in all four quadrants to reflect a shape.
I can rotate 2D shapes about a centre of rotation.

Review questions

1 a These are examples of symmetry from nature.
Write down the number of lines of symmetry for each one.

i butterfly   ii flower   iii starfish   iv snowflake

b Write down the order of rotational symmetry for each one.

2 Write down the order of rotational symmetry for each of these artistic patterns.

a

b

(MR) 3 Write the letter of each shape in the correct space in a copy of the table.
You may use a mirror or tracing paper to help you.

a   b   c   d   e   f

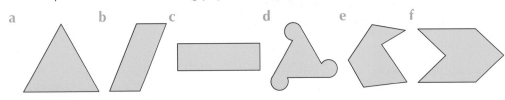

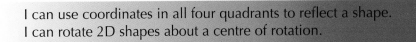

		Number of lines of symmetry			
		0	1	2	3
Order of rotational symmetry	1				
	2				
	3				

4 You can use the rectangular grid and the two L-shapes, below, to make patterns.

This pattern has no lines of symmetry.

Make three copies of the grid.

   a   Use the two L-shapes to make a pattern that has exactly one line of symmetry.

   b   Use the two L-shapes to make a pattern that has exactly two lines of symmetry.

   c   Use the two L-shapes to make a pattern that has rotational symmetry of order one.

(MR) 5 This is an example of a semi-regular tessellation.

You should see that it uses more than one shape.

   a   How many different shapes can you find?

   b   Write down the names of the shapes that you can identify.

   c   How many lines of symmetry does each shape have?

6 Draw a coordinate grid, with both axes numbered from −5 to 5.

Copy this rectangle onto your grid.

   a   Rotate the rectangle ABCD through 90° clockwise about the point (−1, −2), to give the image A'B'C'D'.

   b   Write down the coordinates of the points A', B', C' and D'.

   c   Which coordinate point remains fixed throughout the rotation?

   d   What rotation will move the rectangle A'B'C'D' onto the rectangle ABCD?

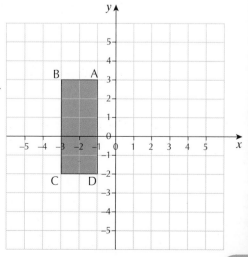

Activity
Rangoli patterns

Design your own rangoli pattern

You will need:

- square dotted paper
- a ruler
- coloured pencils or felt-tip pens.

Work carefully through the steps below to discover how to draw a rangoli pattern.

Step 1

Draw five lines in the first square.

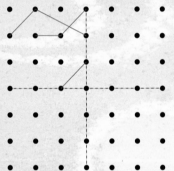

Step 2

Now reflect these lines in the horizontal and vertical lines of symmetry.

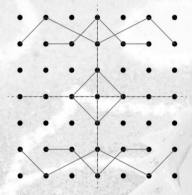

Step 3

Now reflect these lines in both the diagonal lines of symmetry.

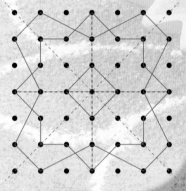

Here is an example of a rangoli pattern drawn on square dotted paper.

Rangoli is a form of folk art from India. Rangoli patterns are decorative designs made on living room and courtyard floors during Hindu festivals. Typically, they are brightly coloured.

The patterns are usually created with materials such as coloured rice, dry flour, sand or even flower petals. The purpose of rangoli is decoration but it is also thought to bring good luck. Designs may also reflect different traditions and folklore.

Rangoli patterns can be simple geometric shapes, representations of Hindu gods or flower and petal shapes. They may be very elaborate designs crafted by numerous people.

Look at these traditional rangoli patterns.

a How many lines of symmetry does each one have?

b What is the order of rotational symmetry of each one?

1

2

3

4

14

Equations

This chapter is going to show you:
- how to solve simple equations
- how to set up equations to solve simple problems.

You should already know:
- how to write and use expressions
- how to substitute numbers into expressions to work out their value
- how to write and use simple formulae.

About this chapter

Algebra is the branch of mathematics where we use letters to stand for numbers. When you wrote formulae in Chapter 7 you were using algebra. You also used algebra in Chapter 10, when you learnt about graphs.

An understanding of algebra helps you in studying mathematics and many other subjects that use mathematics.

Algebra started developing when the ancient Babylonians first worked out rules for doing calculations.

In the third century AD, a Greek mathematician called Diophantus wrote a book about solving equations, which is what you will be doing in this chapter.

Modern algebra started with the work of the Arabic mathematician Al-Khwarizmi (780–850AD). He called his method *al-jabr*, which is Arabic for 'restoration' or 'completion'. From *al-jabr* we get the English word 'algebra'.

14.1 Finding unknown numbers

Learning objective

- To find missing numbers in simple calculations.

In Chapter 7 you learnt how to use **algebra** to write expressions in which letters represent **unknown numbers**. You are going to use those skills in this chapter.

Nat thinks of a number. He doubles it and then adds 7.

Suppose Nat's number is n.

If you double n you get $2n$. If you then add 7 the result is $2n + 7$.

Now Nat says that the answer is 19.

Can you guess the value of n that makes $2n + 7$ equal to 19?

You should be able to see that it is $n = 6$ because $2 \times 6 + 7 = 19$.

Example 1

Molly thinks of a number.

She subtracts 5 and then multiplies by 3.

The answer is 18.

a Call Molly's number m. Write down an expression to show what Molly did.

b Find the value of m that makes the expression equal to 18.

    a If you subtract 5 from m you get $m - 5$.

       If you multiply this by 3 you get $3(m - 5)$.

       The brackets show that you do the subtraction first.

    b Now you want to find the value of m that makes $3(m - 5) = 18$.

       By guessing, you should see that $m = 11$.

       Check: $11 - 5 = 6$ and $3 \times 6 = 18$.

Example 2

In this number wall, the number in each brick is the sum of the numbers in the two bricks below it.

Work out the values of a, b and c.

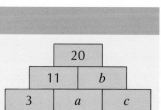

    The top number is always the sum of the two numbers below it.

    $11 + b = 20$, so b must be 9.

    In the same way, $11 = 3 + a$ so $a = 8$.

    Finally, $a + c = b$, so $8 + c = 9$.

    Therefore $c = 1$.

Example 3

Work out the value of the letter x if:

a $x - 3 = 15$ **b** $15 - x = 3$ **c** $\dfrac{x}{3} = 12$ **d** $\dfrac{12}{x} = 3$.

a $x - 3 = 15$

     $x = 18$ x must be 18 because $18 - 3 = 15$.

b $15 - x = 3$ 15 take away 'something' equals 3.

     $x = 12$ x must be 12 because $15 - 12 = 3$.

c $\dfrac{x}{3} = 12$ A number divided by 3 makes 12.

    So $x = 36$.   Because $36 \div 3 = 12$.

d $\dfrac{12}{x} = 3$ What number divides into 12 three times?

    So $x = 4$.   Because $12 \div 4 = 3$.

Exercise 14A

1   Work out the number that each letter represents.

    a $a - 10 = 13$ **b** $b + 14 = 19$ **c** $20 + c = 32$ **d** $23 - d = 1$

    e $40 = e + 29$ **f** $3 = f - 12$ **g** $g + 92 = 100$ **h** $17 = h - 10$

    i $15 = 1 + i$ **j** $14 + j = 30$ **k** $14 - k = 5$ **l** $21 = 16 + l$

2   Work out the number that each letter represents.

    a $3x = 27$ **b** $5y = 20$ **c** $\dfrac{z}{2} = 10$ **d** $40 = 8t$

    e $\dfrac{r}{10} = 5$ **f** $7 = \dfrac{p}{4}$ **g** $3w = 60$ **h** $\dfrac{40}{f} = 20$

    i $\dfrac{100}{a} = 25$ **j** $6j = 66$ **k** $240 = 12k$ **l** $15 = \dfrac{60}{l}$

3   In this number wall, the number in each brick is the sum of the numbers in the two bricks below it.

    Work out the values of a, b and c.

	42	
19		b
10	a	c

4   Here is another number wall.

    Work out the values of g, h and i.

	62	
i		29
g	h	14

5   Work out the number that each letter represents.

    It could be a negative number or a decimal.

    a $2a = 9$ **b** $12 + b = 7$ **c** $c + 6 = 2$ **d** $3 - d = 9$

    e $4e = 22$ **f** $3 = f + 12$ **g** $6g = 3$ **h** $8 = 5 - h$

6   This number wall contains negative numbers.

    Work out the values of a, b and c.

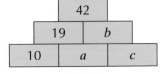

	−9	
−7		a
−3	b	c

7 Rosie thinks of a number. She multiplies it by 3. Then she adds 2.

Call Rosie's number r.

    a Write down an expression to show what Rosie did.

    b Find the value of r, if the answer is 20.

    c Find the value of r, if the answer is 38.

    d Find the value of r, if the answer is 15.5.

  8 George thinks of a number. He adds 8 then multiplies by 2.

Call George's number k.

    a Write down an expression to show what George did.

    b Find the value of k, if the answer is 40.

    c Find the value of k, if the answer is 21.

    d Find the value of k, if the answer is 10.

Reasoning: Bricks in a number wall

A This number wall has four rows and three missing numbers.

Work out the values of a, b and c.

B Make up a number wall of your own, with letters for missing numbers.

You must include at least one number in each row.

Make sure it is possible to solve your number wall.

Give it to someone else to solve.

C What is the largest number of letters you can put in your wall and still make it possible to solve? Justify your answer.

```
                  ┌──────┐
                  │  30  │
              ┌──────┬──────┐
              │  16  │  a   │
          ┌──────┬──────┬──────┐
          │  7   │  9   │  b   │
      ┌──────┬──────┬──────┬──────┐
      │  2   │  5   │  4   │  c   │
      └──────┴──────┴──────┴──────┘
```

14.2 Solving equations

Learning objectives

- To understand what an equation is
- To solve equations involving one operation

Key words

equation	solve

At the start of the previous section Nat was thinking of a number, n.

After multiplying by 2 and adding 7 he got the expression $2n + 7$.

Nat's answer was 19 so you can write $2n + 7 = 19$.

This is an **equation**.

The expressions on each side of the equals sign (=) have the same value.

There is only one value of n that makes $2n + 7$ equal to 19. It is $n = 6$.

Finding the value of n is called **solving** the equation.

Example 4

Solve these equations.

a $x + 17 = 53$ **b** $5y = 45$ **c** $\frac{z}{3} = 9$

a $x + 17 = 53$ Subtract 17 from both sides of the equation.

$\qquad x = 53 - 17$ If you take 17 away from $x + 17$ you are left with x.

$\qquad x = 36$

b $5y = 45$ Divide both sides of the equation by 5.

$\qquad y = \dfrac{45}{5}$ If you divide $5y$ by 5 you are left with y.

$\qquad y = 9$

c $\dfrac{z}{3} = 9$ Multiply both sides by 3.

$\qquad z = 9 \times 3$ If you multiply $\frac{z}{3}$ by 3 you are left with z.

$\qquad z = 27$

Example 5

a Write down an expression for the perimeter of this triangle, in centimetres.

b The perimeter of the triangle is 33.8 cm. Write an equation to express this fact.

c Solve the equation.

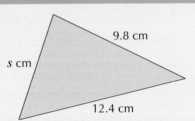

a The perimeter (in cm) $= s + 9.8 + 12.4 = s + 22.2$.

b The perimeter is 33.8 cm so $s + 22.2 = 33.8$.

c Solving the equation means finding the value of s.

$\qquad$ If $s + 22.2 = 33.8$

$\qquad$ then $s = 33.8 - 22.2$

$\qquad\qquad\quad = 11.6$

Sometimes it takes two steps to find the answer.

Example 6

Solve these equations.

a $32 - x = 7$ **b** $\dfrac{24}{y} = 4$

a $32 - x = 7$ Add x to both sides of the equation.

$\qquad 32 = 7 + x$ Now subtract 7 from both sides.

$\qquad$ So $x = 25$.   Be careful! $32 - x = 7$ is not the same as $x - 32 = 7$.

b $\dfrac{24}{y} = 4$ Multiply both sides of the equation by y.

$\qquad 24 = 4y$ Now divide both sides by 4.

$\qquad \dfrac{24}{4} = y$

$\qquad$ So $y = 6$.   Be careful! $\dfrac{24}{y} = 4$ is not the same as $\dfrac{y}{24} = 4$.

Exercise 14B

 1 Solve these equations.

    a $x + 12 = 53$ **b** $y - 12 = 53$ **c** $15 + w = 96$ **d** $t - 2.1 = 3.6$

    e $7.5 = u - 1.3$ **f** $4 = v + 1.2$ **g** $k - 250 = 400$ **h** $90 = 54 + r$

 2 Solve these equations.

    a $3a = 36$ **b** $10b = 35$ **c** $62 = 4c$ **d** $125 = 2k$

    e $\dfrac{r}{4} = 21$ **f** $\dfrac{a}{3} = 30$ **g** $\dfrac{a}{10} = 4.5$ **h** $7.5 = \dfrac{d}{5}$

(PS) **3** **a** Write down an expression for the perimeter of this triangle.

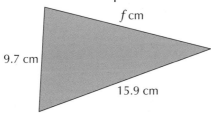

f cm

9.7 cm

15.9 cm

    b Suppose the perimeter of the triangle is 40 cm.

      i Write down an equation involving *f*.

      ii Solve the equation to find the value of *f*.

    c Suppose the perimeter of the triangle is 49.5 cm.

      i Write down an equation involving *f*.

      ii Solve the equation.

(PS) **4** **a** Write down an expression for the area of this rectangle.

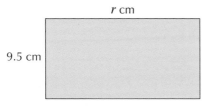

r cm

9.5 cm

    b Suppose the area of the rectangle is 127.3 cm².

      i Write down an equation involving *r*.

      ii Solve the equation.

    c Suppose the area of the rectangle is 96.9 cm².

      i Write down an equation involving *r*.

      ii Solve the equation.

(PS) **5** **a** Look at this quadrilateral.

    The angles of a quadrilateral add up to 360°.

    Use this fact to write down an equation involving *x*.

    b Solve your equation.

    c Find the sizes of the angles of the quadrilateral.

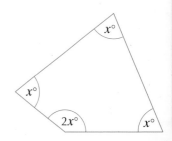

6 For each diagram:

i Write down an equation involving a.   **ii** Solve your equation.

a

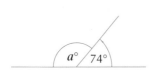

b

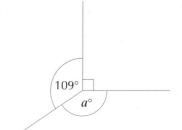

c

d

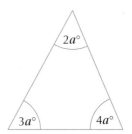

7 Solve these equations.

a $a - 22 = 15$ **b** $22 - a = 15$ **c** $6 = r - 27$ **d** $6 = 27 - r$

e $24 - x = 9$ **f** $1.3 = 5.7 - y$ **g** $70 - t = 31$ **h** $4.25 = 10 - r$

8 Solve these equations.

a $\dfrac{x}{2} = 10$ **b** $\dfrac{x}{10} = 2$ **c** $\dfrac{10}{x} = 2$ **d** $\dfrac{30}{t} = 5$

e $\dfrac{12}{y} = 3$ **f** $\dfrac{100}{k} = 5$ **g** $\dfrac{11}{r} = 2$ **h** $\dfrac{20}{m} = 8$

9 **a** Work out an expression for the number in the top brick of this number wall.

b Suppose the number in the top brick is 80.

   i Write down an equation involving x.

   ii Solve the equation to find the value of x.

c Suppose the number in the top brick is 36.

   i Write down an equation involving x.

   ii Solve the equation to find the value of x.

d Suppose the number in the top brick is 25.

   i Write down an equation involving x.

   ii Solve the equation to find the value of x.

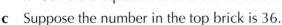

| x | 11 | 14 |

10 **a** Work out an expression for the number in the top brick of this number wall. Write it as simply as possible.

b Suppose the number in the top brick is 11.

   i Write down an equation involving y.

   ii Solve the equation.

c Suppose the number in the top brick is 0.

   i Write down an equation involving y.   **ii** Solve the equation.

| -6 | 5 | y |

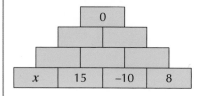

Challenge: Another brick in another wall

Work out the value of *x* in this number wall.

	0		
x	15	−10	8

14.3 Solving more complex equations

Learning objective

* To solve equations involving two operations

Key words

inverse	operation

The equations you solved in the last section involved the **operations** addition, subtraction multiplication or division.

You have seen the equation $2n + 7 = 19$ before. It involves two operations, first 'multiply by 2', and then 'add 7'. You need to 'undo' these operations to work out what *n* is.

* To 'undo' an addition you do a subtraction.

* To 'undo' a multiplication you do a division.

These are **inverse** operations.

Example 7

Solve these equations.

a $4a - 9 = 33$ **b** $3(k + 12) = 81$ **c** $\dfrac{d}{4} - 26 = 14$

a $4a - 9 = 33$

$\quad 4a = 33 + 9$ The inverse of 'subtract 9' is 'add 9' so add 9 to both sides.

$\quad 4a = 42$

$\quad a = 42 \div 4$ The inverse of 'multiply by 4' is 'divide by 4' so divide both sides by 4.

$\quad a = 10.5$

b $3(k + 12) = 81$

In this case, the order of operations is 'add 12' first, then 'multiply by 3'.

When carrying out inverse operations, you need to reverse the original order.

$\quad k + 12 = 27$ The inverse of 'multiply by 3' is 'divide by 3' so divide both sides by 3.

$\quad k = 15$ The inverse of 'add 12' is 'subtract 12' so subtract 12 from both sides.

c $\dfrac{d}{4} - 26 = 14$

$\quad \dfrac{d}{4} = 40$ The inverse of 'subtract 26' is 'add 26' so add 26 to both sides.

$\quad d = 160$ The inverse of 'divide by 4' is 'multiply by 4' so multiply both sides by 4.

Example 8

a Work out an expression involving x for the number in the top brick of this number wall.

b The top number is 20. Write an equation and solve it to find the value of x.

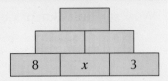

a The expressions on the second row are $x + 8$ and $x + 3$.

The expression in the top box is $x + 8 + x + 3$. This simplifies to $2x + 11$.

b The equation is $2x + 11 = 20$.

$2x = 20 - 11$ First subtract 11 from both sides.

$2x = 9$

$x = \dfrac{9}{2}$ Then divide both sides by 2.

$x = 4.5$

Example 9

In this pentagon, four of the sides are the same length.

a Write an expression for the perimeter of the pentagon, in terms of a.

b Suppose the perimeter of the pentagon is 100 cm. Write down an equation involving a.

c Solve the equation to find the value of a.

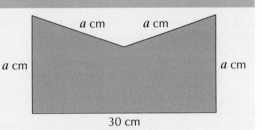

a The perimeter (in cm) is $a + a + a + a + 30 = 4a + 30$.

b If the perimeter is 100 cm, then $4a + 30 = 100$.

c $4a + 30 = 100$

$\quad 4a = 70$ Subtract 30 from both sides.

$\quad\quad a = 17.5$ Divide both sides by 4.

Example 10

Solve these equations.

a $4(a - 3.5) = 23$ **b** $17 - 2k = 5$

a $4(a - 3.5) = 23$

$\quad a - 3.5 = \dfrac{23}{4}$ First divide both sides by 4.

$\quad a - 3.5 = 5.75$ $23 \div 4 = 5.75$

$\quad\quad\quad a = 5.75 + 3.5$ Then add 3.5 to both sides.

$\quad\quad\quad a = 9.25$

b $17 - 2k = 5$

$\quad\quad 17 = 5 + 2k$ First add $2k$ to both sides.

$\quad 17 - 5 = 2k$ Then subtract 5 from both sides.

$\quad\quad 12 = 2k$

$\quad\quad\quad k = 6$ Divide both sides by 2.

Exercise 14C

1 Solve each equation.

a $4x - 5 = 39$ **b** $3y + 7 = 55$ **c** $2t - 13 = 20$ **d** $3 + 4w = 26$

2 Solve each equation.

a $2(a - 5) = 40$ **b** $5(b + 2.5) = 41$ **c** $10(c - 9.1) = 105$ **d** $4(5 + d) = 80$

3 Solve each equation.

a $2(x + 9) = 20$ **b** $2x + 9 = 20$ **c** $3m - 12 = 66$ **d** $3(m - 12) = 66$

e $6y - 11 = 74$ **f** $32 = 2(x - 15)$ **g** $22 = 4x - 7$ **h** $12 + 5x = 43$

i $4(v - 7.5) = 14$ **j** $100 = 5(d + 7)$ **k** $23 = 10 + 2f$ **l** $6(4 + r) = 27$

4 In this hexagon, four of the sides are the same length.

a The perimeter of the hexagon is 150 cm. Write down an equation for this.

b Solve your equation and find the value of x.

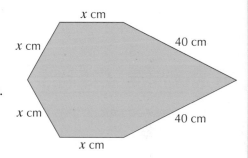

5 In this octagon, six of the sides are the same length.

a Write down an expression for the perimeter of the octagon.

b The perimeter of the hexagon is 2 metres. Write down an equation for this.

c Solve your equation and find the value of y.

d Work out the lengths of the two longest sides of the octagon.

6 Solve each equation.

a $\dfrac{t}{4} = 2$ **b** $\dfrac{k}{4} - 1 = 2$ **c** $\dfrac{t + 2}{3} = 4$ **d** $\dfrac{r - 1}{2} = 10$

e $\dfrac{x}{2} + 17 = 41$ **f** $\dfrac{x + 17}{2} = 41$ **g** $\dfrac{w}{4} - 12 = 15$ **h** $\dfrac{w - 12}{4} = 15$

7 For each of these number walls:

   i use the number in the top brick to write an equation involving x

   ii solve the equation to find the value of x.

a

b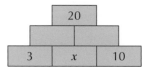

c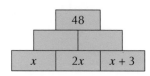

d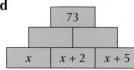

PS **8** Look at this number wall.

Work out the value of x.

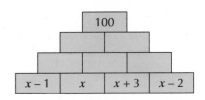

| 100 |
| $x-1$ | x | $x+3$ | $x-2$ |

9 Solve these equations.

a $50 - d = 35$ **b** $2(15 - f) = 20$ **c** $5(8 - y) = 30$ **d** $2(13 - a) = 9$

e $24 = 6(7 - k)$ **f** $9 - 2x = 1$ **g** $31 - 4c = 15$ **h** $25 - 2r = 12$

10 Solve these equations.

a $\dfrac{20}{x} = 4$ **b** $\dfrac{12}{k} - 1 = 2$ **c** $\dfrac{24}{y} + 3 = 7$ **d** $1 + \dfrac{28}{w} = 5$

e $\dfrac{24}{x + 2} = 3$ **f** $\dfrac{30}{x - 1} = 5$ **g** $\dfrac{20}{x + 7} = 2$ **h** $29 = \dfrac{116}{x - 2}$

Challenge: More complex equations

Each of these equations involves more than two steps.

Try to solve them.

a $2(3x + 1) = 66$ **b** $\dfrac{5y - 4}{3} = 17$ **c** $5(3 + 4w) = 65$ **d** $3(33 - 2z) = 60$

14.4 Setting up and solving equations

Learning objective

• To use algebra to set up and solve equations

Many real-life problems can be solved by first setting up an equation.

Example 11

The cost of renting a hostel is £27.90 per night plus £21.50 per person staying.

a The bill for a group of people for one night is £393.40. Write an equation for this, in terms of x, the number of people.

b Solve the equation to find how many people were in the group.

  a The cost in pounds is $21.5x + 27.9$.   This is the number of people $(x) \times 21.5 + 27.9$.

     This could also be written as $27.9 + 21.5x$.

  b The equation is $21.5x + 27.9 = 393.4$.   Do not put a £ sign in the equation.

  c $21.5x + 27.9 = 393.4$

       $21.5x = 365.5$ First subtract 27.9 from both sides. $(393.4 - 27.9 = 365.5)$

          $x = 17$ Then divide both sides by 21.5. $(365.5 \div 21.5 = 17)$

  There are 17 people in the group.

Example 12

Kate thinks of a number.

First she subtracts 13, then she multiplies by 4.

a Call Kate's number k. Write an expression for Kate's answer.

b Write an equation to show this.

c Solve the equation to find Kate's initial number.

My answer is 72.

   a After subtracting 13 from her number, Kate has $k - 13$.

     She then multiplies this by 4 to get $4(k - 13)$.

   b The equationfor Kate's answeris $4(k - 13) = 72$.

   c $4(k - 13) = 72$

$$k - 13 = 18 \qquad \text{First divide both sides by 4.}$$
$$k = 31 \qquad \text{Then add 13 to both sides.}$$

   Kate's initial number is 31.

Exercise 14D

 1 For a club outing, n people go on a coach trip to a castle.

The cost of hiring the coach is £365.

Each person pays £12.75 to go in the castle.

   a Write an expression for the total cost, in pounds, in terms of n.

   b The total cost is £913.25.

     Write an equation involving n and solve it to find how many people went on the trip.

2 Sam thinks of a number, multiplies it by 4 and takes the result away from 50.

   a Use s to stand for Sam's initial number. Write down an expression for the final answer.

   b Work out the value of s, if the final answer is 26.

   c Work out the value of s, if the final answer is 6.

   d Work out the value of s, if the final answer is 13.

3 Mike has m coloured pencils.

Jon has 9 fewer pencils than Mike.

   a Write down an expression, in terms of m, for the total number of pencils.

   b If they have 33 pencils altogether, work out how many each person has.

   c Explain why they cannot have 50 pencils altogether.

4 Josie is 26 years older than her son Tom.

   a Tom is t years old. The total of Tom's and Josie's ages is 104 years. Write an equation, in terms of t, to show this.

   b Solve the equation to find the value of t.

   c How old is Josie?

5 Marie thinks of a number. She adds 15 then multiplies by 2.5.

Suppose Marie's original number is m. Write down an equation in terms of m and solve it.

My answer is 120.

6 This is the plan of the floor of a room. The lengths are in metres.

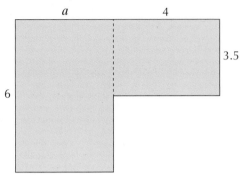

a Suppose the area is 36.5 m². Write an equation to show this and solve the equation to find the value of a.

(PS)

b Suppose the area of the left-hand rectangle is 5.2 m² more than the area of the right-hand rectangle. Write an equation to show this and solve the equation to find the value of a.

7 A formula used in science is $R = \dfrac{E}{I}$.

a If $R = 12$ and $E = 2.4$ then you can write $12 = \dfrac{2.4}{I}$.

Solve this equation to find the value of I.

b Suppose $R = 1.5$ and $E = 10.5$.

Write down an equation and solve it to find the value of I in this case.

8 On Monday r cm of rain fell.

The rainfall on Tuesday was 0.4 cm less than on Monday.

On Wednesday there was twice as much rain as on Tuesday.

a There were 2.8 cm of rain on Wednesday. Write an equation involving r for this.

b Solve the equation to find the value of r.

c How many centimetres of rain fell all together on the three days?

(FS) **9** A group of six friends had a meal in a restaurant. The cost for all six of them was £c.

They decided to add a £10 tip and then share the total equally among the six of them.

a Each person paid £p. Write a formula for p in terms of c.

b If $p = 13.24$, write down an equation involving c.

c Solve your equation to work out the value of c.

10 On Monday, at work, Suzie sent m emails.

On Tuesday she sent one third of the number she sent on Monday.

On Wednesday she sent seven fewer than she did on Tuesday.

a Write down an expression, in terms of m, for the number she sent on Wednesday.

b Suzie sent 19 emails on Wednesday. Write down an equation and use it to work out the value of m.

c How many emails did Suzie send on Tuesday?

11 There are y girls in a school.

The number of girls is 13 less than the number of boys.

There are 967 pupils all together.

a Write down an equation, in terms of y, based on this information.

b Work out the number of girls and the number of boys in the school.

 12 The diagram shows a pond, with a grass border around it. All lengths are in metres.

a Write down an expression for the area, in square metres, of the grass.

b If the area of grass is 35.04 m², write down an equation and solve it to find the width of the pond.

c If the area of grass is 28.8 m², write down an equation and solve it to find the width of the pond.

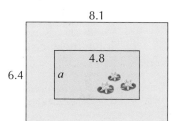

Challenge: Areas

This shape is made out of card. All lengths are in centimetres.

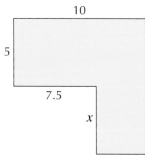

A Show that the total area, in square centimetres (cm²), is $50 + 2.5x$.

B If the area is 64 cm² then $50 + 2.5x = 64$.

Solve this equation to find the value of x.

C Copy and complete this table of values.

Area (cm²)	60	66	72	97	111.5	44
Value of x						

D Show what the shape looks like in the last case, where the area is 44 cm².

Ready to progress?

I can solve simple equations that involve one operation– addition, subtraction, multiplication or division.

I can solve equations that involve two operations, which could be addition or subtraction and multiplication or division.

I can set up and solve an equation for a simple real-life problem.

Review questions

1 Solve these equations.

   a $3.5 + y = 16.2$

   b $4e = 60$

   c $f - 21 = 17$

   d $3z = 25.5$

   e $\dfrac{m}{2} = 22$

   f $\dfrac{r}{3} = 4.5$

   g $5 = t + 8$

   h $23 - w = 7$

   i $\dfrac{18}{x} = 6$

2 a Suppose the volume of this cuboid is 14 cm³.

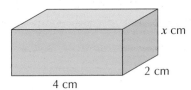

x cm

2 cm

4 cm

   i Write down an equation involving x.

   ii Solve the equation.

   b Suppose the volume of the cuboid is 7.2 cm³.

   i Write down an equation involving x.

   ii Solve the equation.

3 a The angles of a triangle add up to 180°. Write an equation for x.

   b Solve the equation and work out the angles of the triangle.

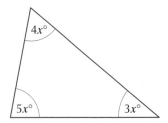

$4x°$

$5x°$

$3x°$

4 Solve these equations.

    a $4a + 17 = 49$ **b** $3(b - 14) = 39$ **c** $5 + \dfrac{c}{4} = 15$

    d $37 = 2d + 15$ **e** $25 - 3e = 4$ **f** $\dfrac{f - 6}{4} = 4.5$

5 A pattern is made from matchsticks.

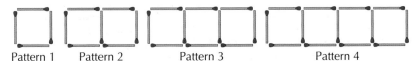

Pattern 1   Pattern 2   Pattern 3   Pattern 4

An expression for the number of matches in pattern p is $3p + 1$.

    a   Show that this expression has the correct value if p is 4.

    b   A particular pattern has 259 matches. Write down an equation involving p to express this.

    c   Solve the equation to work out the value of p.

6 A formula for the mean, m, of two numbers, x and y, is $m = \dfrac{x + y}{2}$.

    a   Put the values $m = 12.7$ and $x = 4.9$ into the formula to give an equation involving y.

    b   Solve the equation to find the value of y.

    c   Use the values $m = 2.1$ and $x = 8.7$ to write down an equation involving y and solve it.

 7 An internet company charges a monthly fee of £12.90 and £1.85 for each film downloaded.

    a   A customer downloads f films in one month. Write down a formula for the cost, c, in pounds.

    b   Mark is charged £36.95 in one month. Write down an equation and solve it to find the number of films he downloaded.

8 I think of a number, add 11.6 and then multiply by 2.4.
The answer is 62.64.

    a   If the original number is n, write down an equation involving n.

    b   Solve the equation to find the original number.

9 **a**   Write down a formula for the perimeter, p cm, of this shape, in terms of y.

    b   If the perimeter of the shape is 23.6 cm, write down an equation for this and solve it to find the value of y.

    c   A formula for the area, A cm², of this shape is $A = 11.25 + 2.5x$.

         If the area is 28 cm², write down an equation for x and solve it.

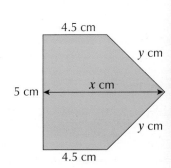

Challenge
Number puzzles

Here is a common type of number puzzle.

Each letter stands for a different number.

The row and column totals are given.

The task is to find the value of each letter.

Using equations can help you do this.

Start by looking for a row or column with just one letter.

A	B	C	A	22
B	C	D	A	23
C	C	C	C	20
D	B	A	D	26
23	16	25	27	

Row 3 has just C. The total is 20 so you can write:

$4C = 20 \rightarrow C = 20 \div 4 = 5$

A	B	5	A	22
B	5	D	A	23
5	5	5	5	20
D	B	A	D	26
23	16	25	27	

Now look for a row or column with just one other letter.

Column 2 has just B.

$2B + 10 = 16$
$\rightarrow 2B = 16 - 10 = 6$
$\rightarrow B = 6 \div 2 = 3$

Now you know B and C.

Row 1 has just one letter.

$2A + 8 = 22$
$\rightarrow 2A = 22 - 8 = 14$
$\rightarrow A = 14 \div 2 = 7$

Column 1 gives $D + 15 = 23$
$\rightarrow D = 23 - 15 = 8$

You should check that all the row and column totals are correct.

A	3	5	A	22
3	5	D	A	23
5	5	5	5	20
D	3	A	D	26
23	16	25	27	

7	3	5	7	22
3	5	D	7	23
5	5	5	5	20
D	3	7	D	26
23	16	25	27	

Solve these number squares. Use equations to help you.

1

A	A	A	B	22
B	C	A	A	31
C	B	A	C	40
D	D	A	D	29
38	38	8	38	

2

A	B	C	D	35
C	A	C	C	42
D	D	C	B	29
D	B	C	A	35
34	32	40	35	

3

A	B	C	D	51
D	D	D	D	56
A	D	A	C	52
D	B	B	A	45
50	48	51	55	

4

A	B	C	D	22
C	B	A	D	22
D	B	A	D	25
C	C	C	D	17
23	11	26	26	

5

A	A	B	C	D	47
B	A	B	C	C	46
E	E	E	E	E	35
C	C	A	D	D	42
E	E	A	A	E	45
43	47	47	43	35	

6 In this square there is no row with just one letter.

A	B	C	6	16
C	B	A	A	13
B	A	C	D	18
A	A	B	C	13
13	16	12	19	

You need to find a different method.

Look at the first two rows. One has A, B, C and 6 with a total of 16.

$A + B + C + 6 = 16$

The other has A, B, C and A with a total of 13.

$A + B + C + A = 13$

If you subtract one from the other you get $6 - A = 3$ and everything else cancels out.

a Solve this equation to find the value of A.

b Solve the number square.

7 Solve this number square.

A	B	C	D	40
D	A	C	B	40
D	B	9	A	38
D	C	B	B	41
49	34	39	37	

8 Solve this number square.

A	B	A	B	38
C	D	A	D	28
A	D	C	B	35
B	5	B	A	34
41	21	41	32	

9 Now make your own puzzle. Make sure it is possible to solve it. Give it to someone else to solve.

15

Interpreting data

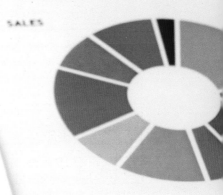

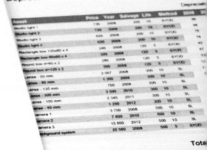

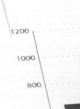

This chapter is going to show you:

- how to use a scaling method to draw pie charts
- how to read and interpret data from a pie chart
- how to use the averages and the range to compare sets of data
- how to carry out and interpret a statistical survey.

You should already know:

- how to draw a simple chart
- how to work out the mode, median, mean and range of a set of data
- how to draw tally charts.

About this chapter

The daily news is full of statistical data about everyone's lifestyles, society and the world around them.

How can you make sense of it all?

One of the most important ways is to use different types of charts and graphs to organise and interpret data and to understand what they can – and cannot – tell you.

To analyse data, you have to select the right tools and know what they can do. Just as importantly, you need to assess what the graphs and charts other people present to you really mean and whether they interpret information in the most useful way, or lead you to false conclusions.

15.1 Pie charts

Learning objectives

- To use a scaling method to draw a pie chart
- To read and interpret data from pie charts

Key words

frequency table	pie chart
scaling	sector

Sometimes you will have to draw a **pie chart** to display data that is given in a **frequency table**. You will also have to interpret pie charts that are already drawn.

Example 1

Draw and label a pie chart to represent this set of data, which shows how a group of people travel to work.

Type of travel	Walk	Car	Bus	Train	Cycle
Frequency	10	20	24	6	12

Each frequency (walk, car, bus and so on) will be represented by a **sector** in the pie chart. The total frequency is 72 people.

A pie chart has an angle of 360° at its centre. To find out the angle that represents one person, you need to divide 360 by 72. This gives an angle of 5° for each person. So, to work out the angle for each sector, multiply the number of people in it (the frequency) by 5.

This is called the **scaling** method; here, the scaling is 5. Multiplying all the frequencies by 5 gives sectors with angles that total 360°.

It is helpful to set the data out in a vertical table and add your calculations to it, like this.

Type of travel	Frequency	Calculation	Angle
Walk	10	10 × 5	50°
Car	20	20 × 5	100°
Bus	24	24 × 5	120°
Train	6	6 × 5	30°
Cycle	12	12 × 5	60°
Total	72	72 × 5	360°

Now draw the pie chart.

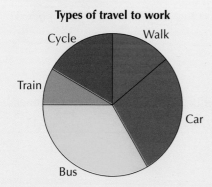

Types of travel to work

Example 2

The pie chart shows the types of housing on a new estate. Altogether there were 540 new houses built.

How many are:

a detached   **b** semi-detached

b bungalows   **d** terraced?

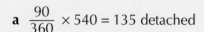

Types of housing

You need to work out the fraction of 540 that each sector represents.

To do this, write the angle of the sector as a fraction of the total angle (360°) in the pie chart.

This will give the fraction of the total number of houses, so multiply it by 540.

a $\dfrac{90}{360} \times 540 = 135$ detached   **b** $\dfrac{120}{360} \times 540 = 180$ semi-detached

c $\dfrac{40}{360} \times 540 = 60$ bungalows   **d** $\dfrac{110}{360} \times 540 = 165$ terraced

Exercise 15A

1 Draw a fully labelled pie chart to represent each set of data.

 a The favourite subject of 36 pupils

Subject	Maths	English	Science	Languages	Other
Frequency	12	7	8	4	5

 b The types of food that 40 people usually eat for breakfast

Food	Cereal	Toast	Fruit	Cooked	Other	None
Frequency	11	8	6	9	2	4

 c The numbers of goals scored by an ice hockey team in 24 matches

Goals	0	1	2	3	4	5 or more
Frequency	3	4	7	5	4	1

 d The favourite colours of 60 Year 7 pupils

Colour	Red	Green	Blue	Yellow	Other
Frequency	17	8	21	3	11

2 The pie chart shows the results of a survey of 216 children about their favourite foods.

How many chose:

a chips   **b** burgers

c fish fingers   **d** curry?

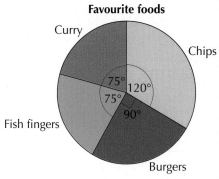

3 The pie chart shows the number of kilograms of butter sold at a corner shop in one week. The total amount sold in the week was 450 kg.

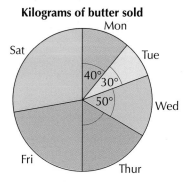

a How many kilograms of butter were sold on Monday?

b If 100 kg was sold on Friday, what is the angle of the sector for Friday?

c One-sixth of all the butter was sold on Thursday. What was the angle of the sector for Thursday?

d What percentage of the week's sales of butter was sold on Saturday?

e On which day was the most butter sold? Give a reason why that might have been.

Activity: Using IT to produce pie charts

Use a spreadsheet for this activity.

A Put the data for each part of question 1 of Exercise 15A into the spreadsheet.

Use a different sheet for each part.

B Use the spreadsheet to produce a pie chart for each set of data.

Ensure that each pie chart has a title and is fully labelled.

C Now use your own data to produce a pie chart about pupils in your class.

For example, you could find out:

* the numbers of boys and girls
* the numbers of left-handed and right-handed pupils
* the numbers with different coloured hair.

15.2 Comparing range and averages of data

Learning objectives

- To use averages and range to compare data
- To make sensible decisions by comparing averages and ranges of two sets of data

Key words

data	mean
median	modal class
mode	range

In Chapter 6, you worked out the **mode**, **median**, **mean** and **range** of sets of **data**. Here is a reminder of how to calculate these, and the advantages and disadvantages of each one. When you use an average to compare two sets of data, it is important to choose the one that will help you find out what you want to know.

	Advantages	Disadvantages	Example
Mean	Uses every piece of data. Probably the most commonly used average.	May not be representative if the data contains extreme values.	2, 2, 2, 2, 4, 12 Mean $= \frac{2 + 2 + 2 + 2 + 4 + 12}{6} = 4$ which is a higher value than most of the data.
Median	The middle value, so it is a better average to use if the data contains extreme values.	Not all values are considered. The median might not be a central value and so could be misleading.	1, 1, 3, 5, 10, 15, 20 Median = fourth value = 5 Note that above the median the numbers are a long way from the median but below the median they are very close.
Mode	The most commonly occurring value.	If the mode is an extreme value it may be misleading to use it as an average.	Weekly wages of a manager and his four staff: £150, £150, £150, £150, £1000 Mode is £150 but mean is £320.
Modal class for continuous data	The largest class in a frequency table. For continuous data the actual values could all be different so the information is clearer when grouped together.	The actual values may not be centrally placed in the class.	**Time (T) minutes** **Frequency** $\quad 0 < T \leqslant 5 \qquad\qquad 2$ $\quad 5 < T \leqslant 10 \qquad\quad 3$ $\quad 10 < T \leqslant 15 \qquad\quad 6$ $\quad 15 < T \leqslant 20 \qquad\quad 1$ The modal class is $10 < T \leqslant 15$, but all six values may be close to 15.
Range	Shows how the data is spread out.	It only looks at the two extreme values.	1, 2, 5, 7, 9, 40 The range is $40 - 1 = 39$. Without the last value (40) the range would be only 8.

Example 3

Which averages could you use to compare the heights of boys and girls in your class?

Give a reason for your answer.

All the heights will probably be different so the mode would not be useful.

The median would give the middle height when the data is written in order, so that could be used.

The mean uses all the heights so that could be used.

Example 4

Three players are hoping to be chosen for the basketball team.

This table shows their scores in the last five games played.

Matt	13	5	12	5	11
Jon	18	6	12	6	10
Sita	15	2	15	6	4

The coach said she would chose according to who had the best average score.

Who would she choose, based on each average?

Put the results into a table.

	Matt	Jon	Sita
Mean	9.2	10.4	8.4
Median	11	10	6
Mode	5	6	15

Mean: She would choose Jon.

Median: She would choose Matt.

Mode: She would choose Sita.

Example 5

A teacher thinks that the girls in her class have quicker reaction times than the boys.

There are 10 boys in the class. Their reaction times, in seconds were:

0.5	0.4	0.3	0.4	0.6	0.3	0.5	0.9	0.6	0.5

There are 12 girls in the class. Their reaction times, in seconds were:

0.2	0.4	0.4	0.6	0.5	0.3	0.4	0.2	0.3	1.0	0.2	0.3

Is she correct? Give reasons for your answer.

The mean for the boys is 0.5 seconds and the range is $0.9 - 0.3 = 0.6$ seconds.

The mean for the girls is 0.4 and the range is $1.0 - 0.2 = 0.8$ seconds.

Using these results the teacher is correct because on average the girls have shorter reaction times than the boys.

However, looking at the range of the times, the boys are more consistent.

1 Look at each set of data and say whether or not the chosen average is a suitable way to interpret it. Explain your answer.

a   2, 3, 5, 7, 8, 10   Mean   b   0, 1, 2, 2, 2, 4, 6   Mode
c   1, 4, 7, 8, 10, 11, 12   Median   d   2, 3, 6, 7, 10, 10, 10   Mode
e   2, 2, 2, 2, 4, 6, 8   Median   f   1, 2, 4, 6, 9, 30   Mean

2 Look at each set of data and decide which average is the most suitable. Explain your answer.

a   1, 2, 4, 7, 9, 10   b   1, 10, 10, 10, 10   c   1, 1, 1, 2, 10
d   1, 3, 5, 6, 7, 10   e   1, 1, 1, 7, 10, 10, 10   f   2, 5, 8, 10, 14

3 Three players hope to be chosen as a striker in their football team.

This table shows the numbers of goals they scored in the last six games they played.

James	0	1	0	4	0	6
Martin	3	0	0	3	4	0
Dan	0	1	0	1	1	1

The coach said she would choose, according to who had the best average score.

Who would be chosen, based on each average?

4 You have to catch a bus regularly. You can catch bus A or bus B. On the last 10 times you caught these buses, you noted down, in minutes, how late they were.

Bus A	3	0	5	11	4	0	4	7	1	10
Bus B	5	6	6	5	1	5	3	6	5	8

By comparing an average and the range, decide which bus you should catch.

Give a reason for your answer.

(MR) 5 You have to choose someone for a quiz team. These are the last ten quiz scores (out of 20) for Alan and Clare.

Alan	1.5	19	2.5	11.5	20	12.5	2.5	6	5	9.5
Clare	8.5	7	9.5	12.5	13	7.5	7	10.5	6.5	8

Who would you choose for the quiz team? Give a reason for your answer.

(PS) 6 Grace and Harry take the same three tests.

The table gives information about their scores.

	Mean	Range	Best score
Grace	17	7	20
Harry	14.0	10	19

a   The teacher wants to give Grace and Harry some feedback.

Use the information to compare their scores.

b   Can you work out their scores on each test?

Challenge: Average parcels

A The modal mass of seven parcels was 400 g.

When a new parcel was added, there were two modes.

State, if possible, the mass of the new parcel.

Explain your answer.

B The median mass of seven parcels was 400 g.

When a new parcel was added, the median changed to 450 g.

Find, if possible, the mass of the new parcel.

Explain your answer.

C The mean mass of seven parcels was 400 g.

When a new parcel was added, the mean changed to 415 g.

Calculate, if possible, the mass of the new parcel.

Explain your answer.

15.3 Statistical surveys

Learning objectives

- To carry out a statistical survey
- To use charts and diagrams to interpret data and then write a report

Key words	
frequency	hypotheses
questionnaire	statistical survey
tally	

When you want to carry out a **statistical survey**, the first thing to do is to choose a problem to investigate. Then:

- decide what type of data you need and how you will collect it
- write down any statements that you want to test in your survey and the answers you might expect. These will be your **hypotheses**.

There are several ways to collect data.

- Carry out a survey of a sample of people. Your sample size should be more than 30. Use a data-collection sheet or a **questionnaire** to collect data from your sample.
- Search secondary sources, such as reference books, newspapers, ICT databases and the internet.

If you use a data-collection sheet, remember these tips.

- Design the layout of your sheet before you start the survey.
- Keep a **tally** of your results.

If you use a questionnaire, remember these tips.

- Make the questions short and simple.
- Use questions that require simple responses; for example: yes, no, do not know or a choice of answers in a tick box.
- Avoid personal and embarrassing questions or leading questions designed to get a particular response.

When you have collected all your data, use statistics, such as the mean, the median, the mode and the range, to interpret (analyse) and compare it.

Then write a report on your findings. Your report should:

• be based only on the evidence you have collected

• use statistical diagrams to illustrate your results and to make them easier to understand

• give reasons why you have chosen to use a particular type of diagram.

To give your report a more professional look, use ICT software to illustrate the data.

Finally, write a short conclusion based on your evidence and your analysis of it.

You may want to refer to your original hypotheses.

Exercise 15C

Write your own statistical report on one or more of these topics. The data can be collected from people in your class or year group, but it may be possible to collect the data from other sources, friends and family outside school or the internet.

1 The amount of TV young people watch

2 The types of sport young people take part in outside school

3 The musical likes and dislikes of Year 7 pupils

4 Investigate this statement.

Smaller people have smaller feet.

5 'More people are taking holidays in the UK this year.' Investigate this statement.

6 'Fewer people are going to live football matches now.' Investigate this statement

7 The number of pages in novels in the school library

8 The price of bars of chocolate

Activity: Write your own report

Write your own statistical report on one or more of these problems. For these problems, you may need to use secondary sources to collect the data.

A Compare the **frequency** of different letters in the English language to the frequency of the same letters in the French language.

B Compare the prices of second-hand cars. Use different motoring magazines.

C Compare the cost of theatre tickets in the UK.

D Compare the number of people in cars on the road at two very different times of the day, for example, 9:00 am and 12 noon.

Ready to progress?

I can use a scaling method to draw pie charts.
I can use the angles of sectors to interpret data from pie charts.
I can use a suitable average (mean, mode or median) and the range to compare data.
I can conduct a statistical survey, interpret the data and write a report.

Review questions

1   Rory and Justin play 18 holes of crazy golf.

These are their scores on each hole.

Rory	5	2	4	4	5	3	4	2	4
	5	3	4	4	4	3	7	4	4
Justin	3	5	4	5	5	4	6	3	6
	5	4	5	4	5	5	4	6	4

  a   What is the modal score for each player?
  b   What is the range of scores for each player?
  c   What is the mean score for each player?

  MR
  d   Which player is more consistent?
      Why do you think so?

  MR
  e   Who is the better player?
      Why do you think so?

2   120 people were asked to name their favourite drink.

  • Three-tenths said tea.

  • 22 people said coffee.

  • 35% of them said cola.

  • The rest said water.

  Show this information in a pie chart.

  Remember to give your chart a title.

PS 3   Compare how long boys and girls usually spend on homework each week.

  a   State a hypothesis.
  b   Collect data.
  c   Interpret your data.
  d   Present a report, including a conclusion.

4 In a survey, a shopkeeper asked 72 customers about the quality of customer service over the previous year. These are the results.

Quality of customer service

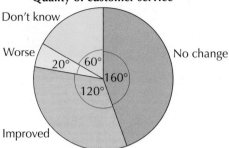

a What percentage of those asked said that customer service was worse?

b How many customers said there had been no change?

c What is the probability that one of these customers chosen at random said customer service had improved?

Give your answers as fractions in their simplest form.

5 There are 1200 pupils in a school.

540 of them are girls.

a What percentage of the pupils are boys?

 b One-third of the girls wear glasses.

200 of the boys wear glasses.

Draw a four-sector pie chart to show all this information.

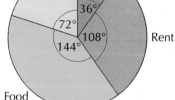 **6** The pie chart shows how Huan spent her weekly pay.

Weekly spending

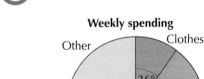

a She spent £45 on clothes each week.

How much did she spend on rent?

b She increased the amount she spent on food each week by 5% and reduced the amount she spent on clothes by 5%. Her rent was unchanged.

Did the proportion she spent on other things increase, decrease or stay the same?

Show your working.

Challenge

Ice-skate dancing competition

In an ice-skate dancing competition there are six couples.

Each week the couples are given scores by three judges.

Each judge gives a score out of 10.

Couple	Week 1	Week 2	Week 3	Week 4	Week 5	Judge
Amos and Gill	5.0	5.0	4.5	7.5	8.5	X
	5.0	6.0	5.5	8.0	9.0	Y
	5.5	6.5	6.0	9.0	9.0	Z
Bernie and Husha	6.0	6.0	6.0	6.5	6.0	X
	6.5	6.5	7.5	6.5	7.5	Y
	6.5	6.5	7.0	6.5	7.5	Z
Carl and Ida	2.0	4.0	3.0	3.0	2.5	X
	3.0	4.5	3.5	4.0	4.5	Y
	3.5	4.5	3.0	4.0	4.0	Z
David and Joanna	5.0	6.0	7.5	8.0	5.5	X
	6.0	7.0	7.5	8.0	6.0	Y
	5.5	7.0	7.5	8.5	5.5	Z
Errol and Kirsty	3.0	3.5	3.0	3.0	3.0	X
	3.5	3.5	4.0	4.5	4.0	Y
	3.5	4.0	4.0	4.0	4.0	Z
Francis and Leela	4.5	3.5	3.5	4.0	7.0	X
	4.5	4.5	4.5	5.5	7.0	Y
	4.5	4.0	4.5	5.0	7.0	Z

Task

Write a newspaper story about the performances of the six couples.

You must include:

- the total scores for each week
- a graph or chart to show the data for all five weeks
- statements to compare the performances over the five weeks
- comments about the judges. Is anyone harsh?
- a prediction about the winner of the competition backed up with evidence
- a headline.

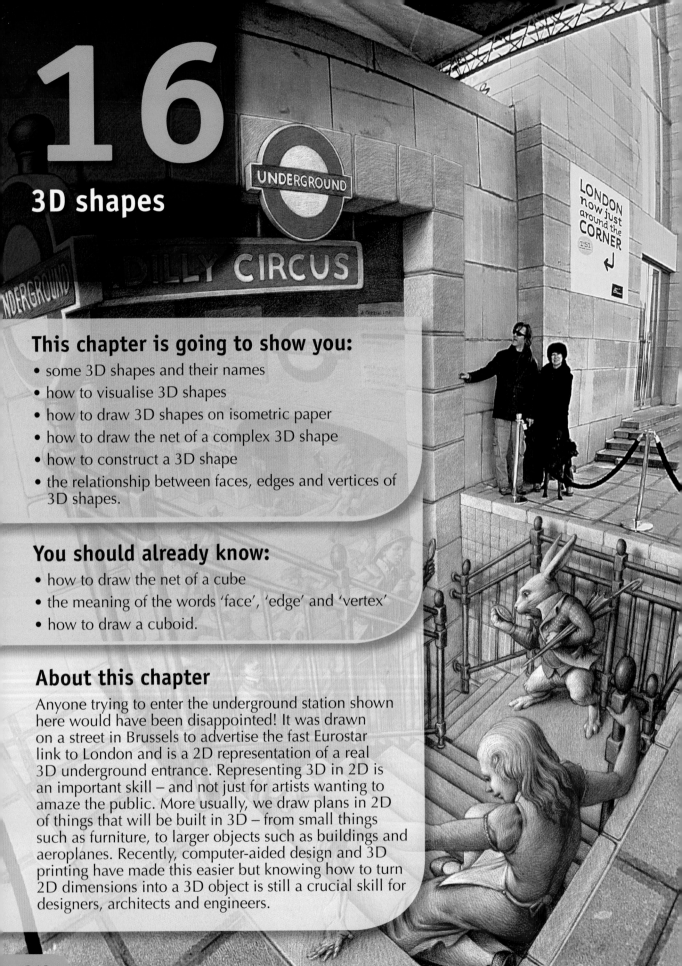

16

3D shapes

This chapter is going to show you:

- some 3D shapes and their names
- how to visualise 3D shapes
- how to draw 3D shapes on isometric paper
- how to draw the net of a complex 3D shape
- how to construct a 3D shape
- the relationship between faces, edges and vertices of 3D shapes.

You should already know:

- how to draw the net of a cube
- the meaning of the words 'face', 'edge' and 'vertex'
- how to draw a cuboid.

About this chapter

Anyone trying to enter the underground station shown here would have been disappointed! It was drawn on a street in Brussels to advertise the fast Eurostar link to London and is a 2D representation of a real 3D underground entrance. Representing 3D in 2D is an important skill – and not just for artists wanting to amaze the public. More usually, we draw plans in 2D of things that will be built in 3D – from small things such as furniture, to larger objects such as buildings and aeroplanes. Recently, computer-aided design and 3D printing have made this easier but knowing how to turn 2D dimensions into a 3D object is still a crucial skill for designers, architects and engineers.

16.1 Naming and drawing 3D shapes

Learning objectives

- To be familiar with the names of 3D shapes and their properties
- To use isometric paper to draw shapes made from cubes

Key words

3D	cuboid
edge	face
hexagonal prism	isometric
pentagonal prism	tetrahedron
triangular prism	vertex

Many everyday objects are made with familiar mathematical shapes. These have three dimensions: length, width and height, so they are called three-dimensional (**3D**).

You need to be able to recognise and name these 3D shapes.

 Cube  **Cuboid**  Square-based pyramid   **Tetrahedron**  **Triangular prism** Cone

 Cylinder   Sphere   Hemisphere 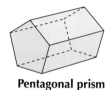 **Pentagonal prism** 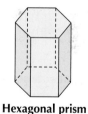 **Hexagonal prism**

3D shapes have **faces**, **edges** and **vertices** (plural of **vertex**). Faces are 2D surfaces, edges occur where two faces meet and vertices are found where three or four faces meet.

Example 1

How many faces, edges and vertices does a square-based pyramid have?

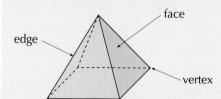

A square-based pyramid has 5 faces, 8 edges and 5 vertices.

You can use **isometric** paper to draw 3D shapes. This paper has dots or lines that form a 60° grid of small triangles.

Example 2

Draw a cube and a cuboid on isometric paper.

Start by drawing a vertical line for the nearest edge and then build the diagram from there.

Draw what you would see.

Hint All of your lines will be either vertical or at 60° to the vertical.

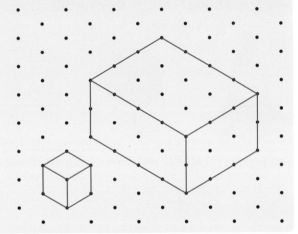

Exercise 16A

1 How many:

i faces   **ii** vertices   **iii** edges

does each 3D shape have?

a   **b**   **c**   **d**

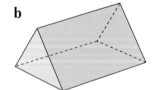

   Cuboid   Triangular prism   Tetrahedron   Hexagonal prism

2 Copy and complete this table. Use the pictures you have seen to help you.

3D shape	Number of faces	Number of edges	Number of vertices
Cube	6	12	8
Cuboid	6	12	8
Square-based pyramid			
Tetrahedron			
Triangular prism	5	9	6
Pentagonal prism			
Hexagonal prism			

3 Explain why cones, cylinders and spheres cannot be described as having a number of 'faces'.

 4 Decide whether each statement is *always true*, *sometimes true* or *never true*.

    a A cuboid has square faces.

    b A triangular prism has five faces.

    c A square-based pyramid has the same number of vertices as edges.

    d A cube is a cuboid.

    e A prism is a cylinder.

    f A hemisphere is half a sphere.

PS **5** This 3D shape is made by putting together a cuboid and a triangular prism.

How many faces, edges and vertices does the shape have?

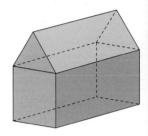

6 Draw each cuboid accurately on an isometric grid.

    a

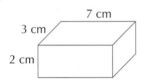
7 cm, 3 cm, 2 cm

    b

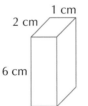
1 cm, 2 cm, 6 cm

    c

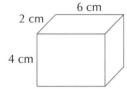
6 cm, 2 cm, 4 cm

7 Use an isometric grid to draw:

    a an L-shape made from four cubes

    b a T-shape made from five cubes

    c a +-shape made from five cubes.

PS **8** How many cubes are required to make this 3D shape?

Use an isometric grid to draw other similar 3D shapes of your own.

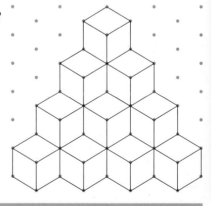

Activity: Cubes to cuboids

You will need 48 unit cubes for this activity.

A Arrange all 48 cubes to form a cuboid.

B How many different cuboids can you make?

16.2 Using nets to construct 3D shapes

Learning objectives

- To draw nets of 3D shapes
- To construct 3D shapes from nets including more complex shapes

Key words

construct

dodecahedron

net

octahedron

truncated square-based pyramid

A **net** is a 2D shape that can be folded to make a 3D shape.

These are nets for a cube, a tetrahedron and a square-based pyramid.

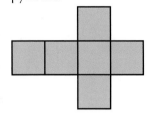 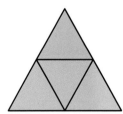 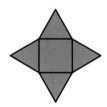

To draw a net and **construct** a 3D shape from it, you will need card, a sharp pencil, a ruler, a protractor, a pair of scissors and glue or tape.

Before folding a net you have drawn, use scissors and a ruler to score the card along the fold lines.

You can fix the edges together by using glue on the tabs or you can just use tape. If you use tabs, keep one face free of tabs and glue this face last.

Example 3

Construct a square-based pyramid.

1   Cut out the net.

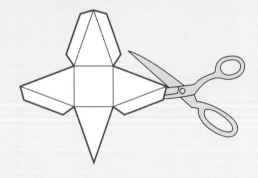

2 Score the fold lines.

3   Fold along the fold lines and glue the tabs.

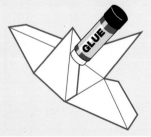

4 Stick down the last face.

For questions 1 to 5, draw the nets accurately on card. Cut out the nets and construct the 3D shapes.

1 A cuboid

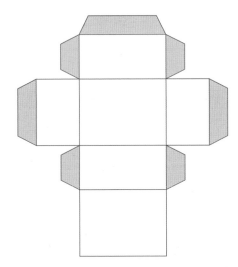

The rectangles have these measurements.

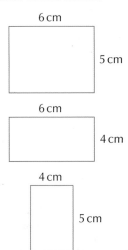

6 cm

5 cm

6 cm

4 cm

4 cm

5 cm

2 A tetrahedron

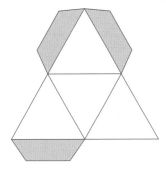

Each equilateral triangle has these measurements.

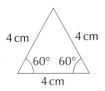

4 cm   4 cm

60° 60°

4 cm

3 A square-based pyramid

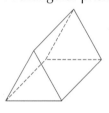

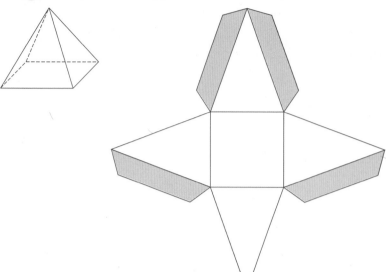

The square has these measurements.

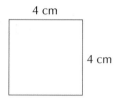

4 cm

4 cm

Each isosceles triangle has these measurements.

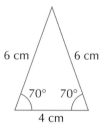

6 cm   6 cm

70° 70°

4 cm

4 A triangular prism

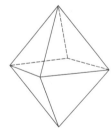

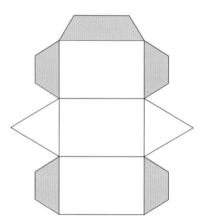

Each rectangle has these measurements.

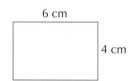

6 cm

4 cm

Each equilateral triangle has these measurements.

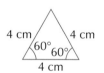

4 cm   4 cm

60° 60°

4 cm

5 An **octahedron**

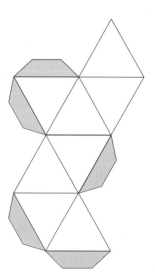

Each equilateral triangle has these measurements.

4 cm   4 cm

60° 60°

4 cm

6 The diagram shows a net for a hexagonal prism.

a Which edge is Tab 1 glued to? On a copy of the diagram, label this A.

b Which edge is Tab 2 glued to? On a copy of the diagram, label this B.

c The vertex marked with a blue star meets two other vertices. Label these with stars.

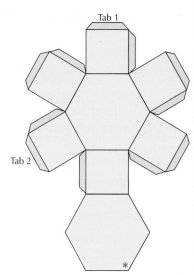

7 Below are the nets for two 3D shapes. What are the two shapes?

a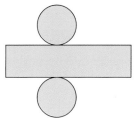

b

Activity: Constructing more complex 3D shapes

The following nets are for more complex 3D shapes. Choose suitable measurements to make each object from card. You may need to add tabs to your net.

A **truncated square- based pyramid**

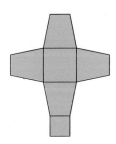

A **dodecahedron**

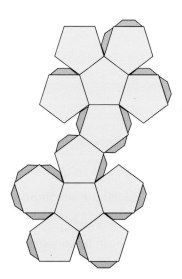

16.3 3D investigations

Learning objectives

- To understand the relationship between faces, edges and vertices for 3D shapes
- To solve problems involving 3D shapes

Key words

| Euler | hexomino |

Leonard **Euler** was a famous eighteenth-century Swiss mathematician. He discovered a relationship between the numbers of faces, edges and vertices of 3D shapes. This relationship does not apply to cylinders, spheres or cones.

Exercise 16C

1 **a** Copy and complete the table.

	Number of faces (F)	Number of edges (E)	Number of vertices (V)
Cube	6	12	8
Cuboid			
Tetrahedron			
Square-based pyramid			
Triangular prism			
Hexagonal prism			
Octahedron			

b Can you spot the relationship between the numbers of faces, edges and vertices?

2 A **hexomino** is a 2D shape made from six squares that touch, side to side. Here are three examples.

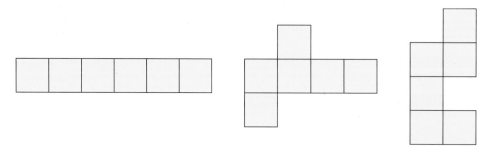

a There are 35 different hexominoes.

Draw, on squared paper, as many different hexominoes as you can.

b How many of these hexominoes are nets for cubes?

On an isometric grid, see how many different 3D shapes you can draw with five cubes. Here is an example.

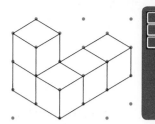

Problem solving: Isometric drawings

Here are two 3D shapes, each made from five cubes.

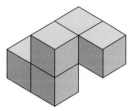 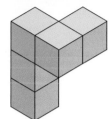

A Copy the diagrams onto isometric paper.

B Now draw the shapes on isometric paper as if they were rotated 90° clockwise.

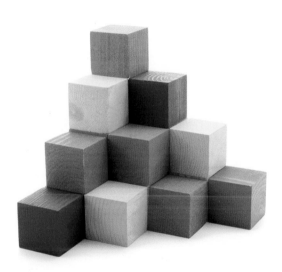

Ready to progress?

I can name 3D shapes.
I can draw nets of 3D shapes.
I can construct 3D shapes from nets.
I can visualise 3D shapes.

I can describe the relationship between the numbers of edges, vertices and faces for a 3D shape.
I can draw 3D shapes on isometric paper.
I can construct complex 3D shapes such as hexagonal prisms, octahedrons and dodecahedrons.

Review questions

1   This net can be made into a hexagonal-based pyramid.

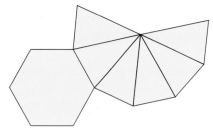

   a   How many faces does the pyramid have?
   b   How many edges does the pyramid have?
   c   How many vertices does the pyramid have?

2   This is a net of a cube.

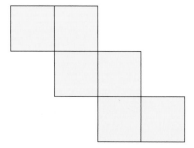

   a   How many tabs are needed to make the cube?
   b   Draw the net and add your tabs.

3 There are 12 spheres, 6 cones, 5 square based pyramids, 4 triangular prisms, 2 cubes and 1 cuboid in a bag.

One of the shapes is picked out at random.

What is the probability that the shape:

a   is a sphere

b   has a curved surface

c   has only 6 vertices?

Give your answers as fractions in their simplest form.

4 A cuboid is made from 48 identical cubes.

The length of the cuboid is double the width.

The height is the same as the sum of the length and the width.

a   Work out the dimensions of the cuboid.

b   Each cube has edges of length 5 cm.
Work out the volume of the cuboid.

5 This is a net of a cuboid.

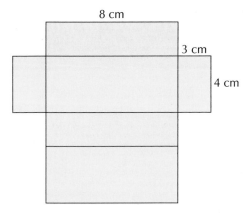

a   Work out the volume of the cuboid.

b   The net is cut from a piece of paper measuring 15 cm × 15 cm.
Work out the percentage of paper that is wasted.

6 a Sketch the net of this triangular prism.

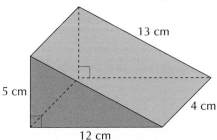

b   Work out the area of the surface of the prism.

c   Show how you can tell that the volume of the prism is 120 cm³.

Problem solving
Packing boxes

Ivan is a delivery driver. He delivers boxes of computer games from a warehouse.

There are two sizes of box.

The large boxes measure 60 cm by 40 cm by 10 cm and contain 192 games.

10 cm

HANDLE WITH CARE

40 cm   60 cm

The small boxes measure 30 cm by 20 cm by 10 cm and contain 40 games.

10 cm

HANDLE WITH CARE

20 cm   30 cm

The back of his van is a cuboid shape measuring 1.8 m by 1.2 m by 1 m.

Task 1

Work out:

a the maximum number of large boxes he can fit in his van

b the maximum number of small boxes he can fit in his van

c the maximum number of boxes he can fit in his van if he has the same number of large and small boxes

d the maximum number of games he can carry in the van when he leaves the warehouse.

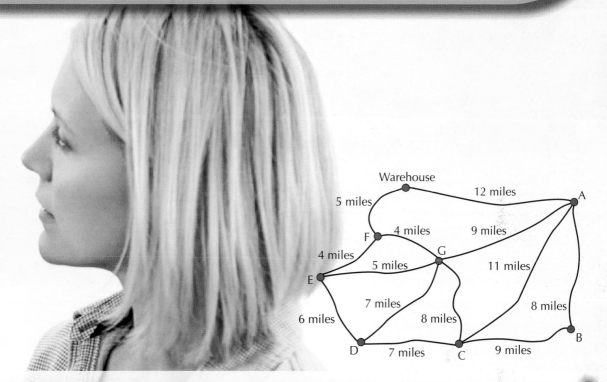

Task 2

Ivan has to drive from the warehouse and back to the warehouse each day.

His delivery route takes him to stores located in seven towns, as shown on the map.

For example: Warehouse → A → B → C → G → D → E → F → Warehouse

On this route he will travel: 12 + 8 + 9 + 8 + 7 + 6 + 4 + 5 = 59 miles

a On one day he delivers to all seven towns but has to visit towns D, E, F and G first.

Write down a possible route he could take.

b On another day he only travels to four of the towns.

He travels 40 miles from the warehouse and back in total.

Work out his route.

Write down the towns he visited in the correct order.

c Work out his shortest route if he visits all seven towns.

17

Ratio

This chapter is going to show you:

- how to use ratio notation with two or three items
- how to use ratios to compare quantities
- how to simplify ratios
- how to write ratios in different ways
- how to use ratios to find totals or missing quantities
- the connection between ratios and fractions
- how to use ratios to solve problems in everyday life.

You should already know:

- how to simplify fractions
- how to work out a fraction of a quantity
- the equivalence between simple fractions and percentages.

About this chapter

The Tour de France is one of the toughest bike races in the world. It lasts 21 days and covers all sorts of terrains, including the high and rugged Alpine mountains. Apart from being enormously fit, how do the riders cope with the steep slopes – and keep up a good speed at the same time? The answer lies in the ratios between the cogs on their bikes – one connected to the pedals and the other on the back wheel. For going uphill they would select a small gear ratio which means that the pedals are linked to a small cog driving a larger one on the wheel. This makes it easier to pedal. For going downhill they would select a large gear ratio. This helps to control the bike.

17.1 Introduction to ratios

Learning objectives

- To use ratio notation
- To use ratios to compare quantities

Key words

quantity	ratio

A **ratio** is a mathematical way to compare **quantities**.

To help you understand what a ratio is, look at the photo of a giraffe and an antelope.

Can you see that the giraffe is about four times as tall as the antelope?

Speaking mathematically, you could say: 'The ratio of the giraffe's height to the antelope's height is four to one.'

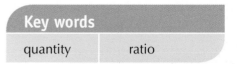

You could also say: 'The ratio of the antelope's height to the giraffe's height is one to four.'

When you need to write a ratio, you need to use the colon (:) symbol. You read this as 'to'.

- The ratio of the giraffe's height to the antelope's height is 4 : 1 (four to one).

- The ratio of the antelope's height to the giraffe's height is 1 : 4 (one to four).

Example 1

The mass of a lion is 156 kg.

The mass of a domestic cat is 4.6 kg.

What is the ratio of the mass of the lion to the mass of the cat?

    Compare the masses of the two animals.

    The lion is heavier than the cat.

    $156 \div 4.6 = 33.91\ldots \approx 34$

    The lion is approximately 34 times as heavy as the cat.

    The ratio of the lion's mass to the cat's mass is 34 : 1.

Example 2

Bottled water comes in two sizes, 500 millilitres (ml) and 750 ml.

What is the ratio of the two sizes?

    The larger bottle is $1\frac{1}{2}$ times as big as the smaller one.

    $1\frac{1}{2} = \frac{3}{2}$

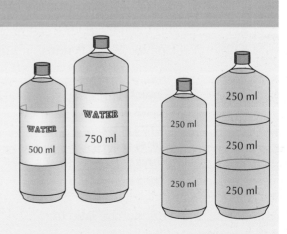

    Imagine each bottle, divided into 250 ml sections.

    There are three sections in the larger bottle and two sections in the smaller one.

    So   the ratio of the capacity of the larger bottle to the capacity of the smaller one is 3 : 2

    or   the ratio of the capacity of the smaller to the larger is 2 : 3.

1 Andy spends £45 and Clare spends £180.

   a Work out the ratio of the amount Andy spends to the amount Clare spends.

   b Work out the ratio of the amount Clare spends to the amount Andy spends.

2 Here are the ingredients to make 12 fruit slices.

   Work out the ratio of:

   a the mass of oats to the mass of sultanas

   b the mass of sugar to the mass of flour

   c the mass of apricots to the mass of sugar.

Flour	100 g
Butter	200 g
Sultanas	100 g
Apricots	250 g
Oats	200 g
Sugar	150 g

3 Shampoo is sold in plastic bottles, in three sizes.

   The small bottle holds 250 ml, the medium holds 375 ml and the large holds 500 ml.

   Work out the ratio of the capacity of:

   a the small to the medium   **b** the medium to the large

   c the small to the large.

4 At a football match 25% of the spectators are female. The rest are male.

   Work out the ratio of males to females.

5 Nadia buys a 2 kg bag of sugar and a 250 g packet of cornflour.

   Work out the ratio of the mass of sugar to the mass of cornflour.

6 The shortest athletics track event is 100 metres and the longest is 10 km.

   Work out the ratio of these two distances.

7 The number of pages in a mathematics book is 80% of the number of pages in a science book.

   Work out the ratio of the numbers of pages in the two books.

8 Liz's age is two-thirds of Debbie's age.
   Debbie's age is half of Freya's age.
   What is the ratio of Liz's age to Freya's age?

9 A recipe for 'Thick Perkins', from 1927, includes these ingredients.

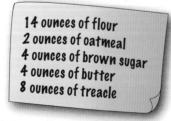

14 ounces of flour
2 ounces of oatmeal
4 ounces of brown sugar
4 ounces of butter
8 ounces of treacle

 Hint An ounce is an old unit of weight.

   a Work out the ratio of:

   i flour to oatmeal   **ii** flour to brown sugar   **iii** oatmeal to butter to treacle.

   b What percentage of all these ingredients is treacle?

   c Work out the ratio of butter to the other ingredients.

10 Engine sizes of cars are measured in millilitres or in litres.

This chart shows the engine sizes of five models of car.

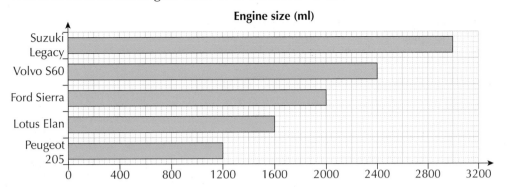

Engine size (ml)

a Work out the ratios of these engine sizes.

   i   Volvo S60 to Peugeot 205   **ii**   Lotus Elan to Ford Sierra

   iii   Suzuki Legacy to Peugeot 205

b The ratio of the engine size of a Dodge Viper to a Lotus Elan is 4 : 1.

   Work out the ratio of the engine size of a Peugeot 205 to the engine size of a Dodge Viper.

11 The average mass of a baby born in England and Wales is 3.4 kg.

The average mass of an adult male in England and Wales is 84 kg.

Show that the ratio of these masses is approximately 25 : 1.

12 A website gives these figures.

Population of UK	63 182 000
Population of China	1 354 040 000

Work out the approximate ratio of the population of China to the population of UK, in the form $N : 1$ where N is a whole number.

13 These are the distances from London to some other capital cities.

Capital city	Paris	Rome	Washington DC	Auckland
Distance (km)	342	1434	5914	18 327

Work out the approximate ratios of these distances, in the form $1 : N$ where N is a whole number.

a London to Paris and London to Rome

b London to Paris and London to Washington DC

c London to Paris and London to Auckland

14 These are the areas of four countries.

Country	United Kingdom	Netherlands	China	Russian Federation
Area (thousands of km²)	244	42	9584	17 075

Work out the approximate ratios of the areas of these countries, in the form $N : 1$ where N is a whole number.

a UK and Netherlands   **b** China and UK   **c** Russian Federation and UK

15 An article in the internet says: 'One out of 10 people are left-handed.'

Simon says: 'That means the ratio of left-handed people to right-handed people is 1 : 10.'

Simon has made a mistake. Correct it for him.

When customers pay cash into a bank, the coins are put into bags.

Each bag holds just one type of coin.

This is the label on a bag.

The number of coins must make up a specified value.

For example, a bag of 20p coins must be worth £10.

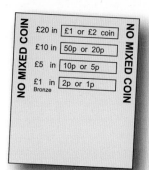

A **a** Work out the ratio of the number of coins in a 20p bag to the number of coins in a 5p bag.

   b Is it the same as the ratio of the values of the two bags?

B Investigate the ratios of the numbers of coins in other bags and compare them to the ratios of the values of the bags.

17.2 Simplifying ratios

Learning objectives

* To write a ratio as simply as possible with whole numbers
* To write ratios in the form $n : 1$ or $1 : n$ where n could be a decimal

Key words

aspect ratio	fraction
simplify	

You have looked at some simple ratios such as 5 : 1 and 3 : 2.

In this section you will look at more complicated ratios. You will find out how to write a ratio as simply as possible. You will also look at an alternative way to write ratios.

Example 3

The heights of two office blocks are 12 metres and 32 metres. What is the ratio of the heights of the office blocks?

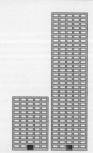

You can write the ratio as 12 : 32.

You can **simplify** a ratio in the same way as you do a **fraction**: divide both numbers by a common factor.

4 is a common factor of 12 and 32.

12 : 32 = 3 : 8

This is the ratio in its simplest form.

This means that:

* the height of the shorter block is $\frac{3}{8}$ of the height of the taller block
* the height of the taller block is $\frac{8}{3}$ or $2\frac{2}{3}$ of the height of the shorter block.

You usually write ratios in the form $A : B$ where A and B are whole numbers.

Sometimes a ratio cannot be simplified to give small whole numbers. In those cases it is helpful to use decimals and write the ratio in the form $1 : n$ or $n : 1$, where n is a decimal.

Example 4

A running club has 214 men and 167 women as members.

Work out the ratio of men to women:

a in the form $n : 1$ **b** in the form $1 : n$.

The ratio of men to women is 214 : 167.

This cannot be simplified to smaller whole numbers.

a Divide both numbers by 167. The ratio of men to women is 1.28 : 1.

b Divide both numbers by 214. The ratio of men to women is 1 : 0.78.

This means that:

there are 1.28 men for every woman, or

there are 0.78 women for every man in the club.

Exercise 17B

Write your ratio answers as simply as possible.

1 Simplify these ratios as much as possible. Give your answers in integers.

   a 5 : 40   **b** 18 : 24   **c** 250 : 50   **d** 480 : 360   **e** 40 : 60

2 In a wood there are 30 birch trees and 105 ash trees.

   a Work out the ratio of birch trees to ash trees.

   b Work out the ratio of ash trees to birch trees.

   c Work out the missing fraction in this sentence.

   The number of birch trees is … of the number of ash trees.

   d Work out the missing number in this sentence.

   The number of ash trees is … times the number of birch trees.

3 A drink is made from 50 ml of squash and 175 ml of water.

   a Work out the ratio of squash to water.

   b Write the amount of squash as a fraction of the amount of water.

4 A display has 1200 coloured lights.

   300 are white, 450 are blue, 200 are green and the rest are yellow.

   Work out the ratio of:

   a white to blue   **b** blue to green   **c** green to yellow.

5 Write each of these ratios in the form $x : 1$, rounding x to one decimal place (1dp).

   a 57 : 17   **b** 100 : 7   **c** 292 : 63   **d** 1267 : 42   **e** 5000 : 3521

6 A council gardener plants these bulbs.

a Work out the ratio of the numbers of:

   i tulip bulbs to iris bulbs

   ii iris bulbs to hyacinth bulbs

   iii hyacinth bulbs to tulip bulbs.

| 120 tulips |
| 300 irises |
| 180 hyacinths |

b Out of the 600 bulbs, 558 flower successfully.

Work out the ratio of flowering to non-flowering bulbs, in the form $x : 1$.

7 The label on a cardigan shows this information.

a Work out the ratio of viscose to cashmere, in the form $x : 1$.

b Work out the ratios for some other pairs of ingredients.

Give your answers in the form $x : 1$.

| 48% viscose |
| 36% polyamide |
| 8% angora |
| 8% cashmere |

8 This table gives the masses, in grams, of one cubic centimetre (1 cm³) of gold and various other metals.

Metal	Gold	Lead	Silver	Copper	Iron	Titanium	Aluminium
Mass (g)	19.3	11.4	10.5	9.0	7.9	4.5	2.7

a Which two metals have masses in the ratio 2 : 1?

b Which two metals have masses approximately in the ratio 4 : 1? Justify your answer.

c Work out the ratio of the masses of gold and aluminium.

9 This table shows the numbers of medals won by some countries in the 2012 Olympics.

Country	GB	Spain	Australia	Germany	New Zealand	Italy
Medals	65	17	35	44	13	28

a Write the ratio of medals for GB to each of the other countries, in the form $x : 1$, where x is a number rounded to one decimal place.

b The USA won more medals than any other country.

The ratio of medals won by the USA to those won by GB was 1.6 : 1.

How many medals did the USA win?

10 A4 and A3 and A2 are three standard sizes of paper.

The sizes are given in this chart.

	Length (mm)	Width (mm)
A4	297	210
A3	420	297
A2	594	420

a Work out the ratio of the length to the width for each size of paper. Write it in the form $n : 1$.

b Work out the ratio of the area of A4 paper to the area of A3 paper.

c Work out the ratio of the area of A4 paper to the area of A2 paper.

d The next size up after A2 is A1. Work out the size of A1 paper.

11 The nutritional information on the pack of sunflower spread lists four different types of fat.

Type of fat	per 100 g
Saturates	15 g
Monounsaturates	19 g
Polyunsaturates	30 g
Omega 3	4 g
Total	68 g

a Work out the ratio of:

    i poyunsaturates to omega 3   ii polyunsaturates to saturates.

b Work out the ratio of polyunsaturates to monounsaturates, giving the answer in the form $x : 1$.

c Work out the ratio of fat to other ingredients in the spread. ·

Activity: Aspect ratios

The **aspect ratio** for a television or cinema screen is the ratio of the width to the height.

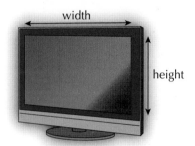

For modern TVs the aspect ratio is 16 : 9.

For older TVs the aspect ratio is 4 : 3.

For widescreen cinema the aspect ratio is 12 : 5.

A For each of those aspect ratios, work out a **fraction** to complete this sentence:

The height of the screen is … of the width.

B Write each of the aspect ratios in the form $x : 1$, where x is a number rounded to one decimal place.

C A laptop screen has a width of 344 mm and a height of 194 mm.

Work out the aspect ratio for this screen.

D Measure any other screens to which you have access.

Work out the aspect ratio of each one.

17.3 Ratios and sharing

Learning objectives

- To use ratios to find totals or missing quantities
- To write ratios to compare more than two items

So far, you have used ratios to compare two different quantities. You can also use ratios for sharing out a given quantity into different amounts.

Example 5

Two charities share a donation of £20 000 in the ratio 1 : 4. How much does each charity get?

The total number of shares is $1 + 4 = 5$.

Imagine that £20 000 is divided into five equal parts.

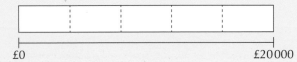

£0   £20 000

The ratio means that one charity gets one part or $\frac{1}{5}$ of the money and the other charity gets

four parts or $\frac{4}{5}$.

$\frac{1}{5}$ of 20 000 = 4000 $\frac{4}{5}$ of 20 000 = $4 \times 4000 = 16\,000$

One charity gets £4000 and the other gets £16 000.

In Example 5 you knew the total amount.

If you know one of the shares, you can use this to work out the total amount.

Sometimes, ratios have more than two parts.

Example 6

Robena makes a drink by mixing orange juice, pineapple juice and lemonade in the ratio 3 : 6 : 1.
She uses 150 ml of orange juice.

How much of the drink will there be altogether, if Robena follows her usual recipe?

The number of parts is $3 + 6 + 1 = 10$.

Orange juice is $\frac{3}{10}$ of the drink, pineapple is $\frac{6}{10}$ and lemonade is $\frac{1}{10}$ of the drink.

$\frac{3}{10} = 150$ ml $\rightarrow \frac{1}{10} = 150 \div 3 = 50$ ml $\rightarrow \frac{10}{10} = 50 \times 10 = 500$ ml

She makes 500 ml altogether.

Exercise 17C

1 In a class of 35 children, the ratio of swimmers to non-swimmers is 6 : 1.
How many of the children are swimmers?

2 Two sisters, Helga and Martina, shared the cost of a computer game in the ratio 3 : 2.
The game cost £39.50.
How much did each sister pay?

3 A farmer has cows, sheep and goats.
The ratio of cows to sheep to goats is 6 : 3 : 1.
  a What percentage of the animals are cows?
  b There are 18 sheep.
    Work out the numbers of cows and goats.

4 In a small library, the ratio of fiction to non-fiction books is 1 : 8.
There are 960 non-fiction books.
How many books are there altogether?

5 Harriet, Richard and Steve collect postcards. The ratio of the numbers of cards they have is 4 : 5 : 6.
They have 240 cards altogether.
How many cards does each person have?

 6 At a concert the number of men to women is in the ratio 1.5 : 1. There are 150 people altogether.
How many women are at the concert?

7 400 people see a film at a cinema. The numbers of adults, teenagers and children are in the ratio 1 : 2 : 5.
Show that 250 children see the film.

8 A bakery bakes white, granary and wholemeal loaves in the ratio 4 : 4 : 1.
The bakery bakes 180 white loaves.
How many loaves are baked all together?

9 In a fishing contest the numbers of trout and carp caught were in the ratio 1 : 2.5.
The total number of trout was 26.
How many carp were caught?

10 The ratio of teachers to students in a school is 1 : 12.6.

    a There are 83 teachers.

      Work out the number of students.

    b In another school the ratio of teachers to students is 1 : 10.8.

      There are 486 students in this school. How many teachers are there?

11 In 2012, in a particular town, over a period of 60 days each day was classified as sunny, cloudy or wet. The ratio of sunny to cloudy to wet days was 7 : 3 : 2.

    a How many days were sunny?   **b** What percentage of the days were cloudy?

 12 A football pitch is 95 metres long. The ratio of the length to the width is 2.5 : 1.

    Work out the perimeter of the pitch.

Challenge: Mixing gold

Gold is used to make jewellery.

Gold is a soft metal and it will gradually wear away in everyday use.

To make it harder it is mixed with other metals.

The other metals used include copper, silver, nickel, palladium and zinc.

Gold is very expensive, so mixing it with other metals also makes the jewellery less expensive.

The purity of gold is described in carats.

    Pure gold is 24 carats.

    21 carat gold is 87.5% gold and the rest is other metals.

    18 carat gold is 75% gold and the rest is other metals.

    9 carat gold is 37.5% gold and the rest is other metals.

A Work out the ratio of gold to other metals in:

    a 18 carat gold   **b** 21 carat gold   **c** 9 carat gold.

B A jeweller has 5.00 g of pure gold.

    a How much 9 carat gold can she make from this?

    b How much 18 carat gold can she make from this?

    c How much 21 carat gold can she make from this?

C It is also possible to buy 22 carat gold.

    a What is the percentage of pure gold in 22 carat gold?

    b Work out the ratio of gold to other metals in 22 carat gold.

D Use a ratio to compare the amount of pure gold in equal masses of 9 carat gold and 18 carat gold.

E The cost of gold varies from day to day. Try to find out the cost of one gram of gold.

17.4 Solving problems

Learning objectives

- To understand the connection between fractions and ratios
- To understand how ratios can be useful in everyday life

Ratios are useful in practical situations.

Example 7

There is room for 150 students and teachers on a school trip. The teacher-to-student ratio must be 1 : 8 or better.

What is the smallest possible number of teachers that can go on the trip?

If the ratio of teachers to students is 1 : 8 then the students make up $\frac{8}{9}$ of the group and the teachers make up $\frac{1}{9}$ of the total number.

$\frac{1}{9}$ of the total is $150 \div 9 = 16.67$.

The smallest possible number of teachers is 17. (You cannot take 0.67 of a teacher!)

Exercise 17D

1 There are 450 pupils in a primary school.

The ratio of boys to girls is 5 : 4.

Work out the numbers of boys and girls.

2 Ahmed has downloaded some music tracks. 57% of them are dance tracks.

Work out the ratio of dance tracks to other music.

Give your answer in the form $n : 1$.

3 On a bus one Tuesday morning, two-thirds of the passengers are children on the way to school.

The rest of the passengers are equal numbers of men and women.

Work out the ratio of schoolchildren to men to women.

 4 James is saving 5p and 10p coins. He has 75 coins.

The ratio of 5p to 10p coins is 1 : 1.5.

How much are his coins worth?

 5 There are 30 green bottles and brown bottles on a wall.

The ratio of green bottles to brown bottles is 1 : 4.

a One of the green bottles accidentally falls.

What is the ratio of green bottles to brown bottles now?

b Now one of the brown bottles falls.

What is the ratio of green bottles to brown bottles now?

6 **a** The ratio of Hassan's age to his mother's age is 1 : 1.64.

The total of their ages is 66 years.

What are their ages?

7 This pie chart shows the area of each country in the United Kingdom.

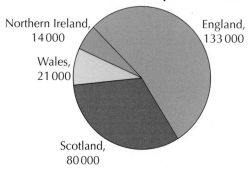

Areas of UK countries in square kilometres

Northern Ireland, 14 000

England, 133 000

Wales, 21 000

Scotland, 80 000

a Use ratios to compare the areas of England, Scotland and Northern Ireland to the area of Wales.

MR **b** Explain a method for using a ratio to compare the area of any country to the area of Wales.

8 In 2010 there was a general election in the UK.

This table shows the numbers of votes and the numbers of members of parliament (MPs) for the three main parties.

Party	Votes	MPs
Conservative	10 703 654	306
Labour	8 606 515	258
Liberal Democrat	6 836 246	57

a Work out, in the form $x : 1$, the ratio of:

i the number of Conservative votes to the number of Labour votes

ii the number of Conservative MPs to the number of Labour MPs.

b Repeat part **a** but, this time, for the Conservative Party and the Liberal Democrat Party.

MR **c** Some people think that the current system of electing MPs is unfair on parties that get smaller numbers of votes.

Explain whether your answers to parts **a** and **b** give any support to this view.

PS **9** The ratio of Sam's age to his father's age is 1 : 4.5.

The ratio of Sam's age to his mother's age is 1 : 3.5.

Sam's father is 54 years old.

How old is Sam's mother?

PS **10** In this trapezium:

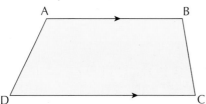

the ratio of angle A to angle D is 3 : 2
the ratio of angle B to angle C is 5 : 4.
Work out the angles of the trapezium.

Investigation: Patterns of squares

Here is a pattern of squares.

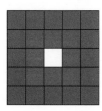

There is a white square in the centre.

It is surrounded by a ring of red squares and then a ring of blue squares.

A Work out the ratio of blue squares to red squares.

B A third ring of squares is added and coloured red.
Work out the new ratio of blue squares to red squares.

C A fourth ring of squares is added and coloured blue.
Work out the ratio of blue squares to red squares now.

D Investigate how the ratio will change if you continue to add red and blue rings alternately.

E Can you work out a way to predict what the ratio of blue to red will be if there is a large number of rings, such as 10 or 20 or 30?

Ready to progress?

Review questions

(PS) 1 A lawn fertiliser recipe uses ammonium sulphate, iron sulphate and sharp sand in the ratio 3 : 1 : 20. 100 grams of fertiliser are needed for each square metre of lawn.

    a How much ammonium sulphate is needed to make 100 g of fertiliser?

    b Dave's lawn is a rectangle, 4 m long and 9 m wide. How much sharp sand does he need to fertilise his lawn?

2 a Write down the following ratios for a cube, as simply as possible.

      i The number of edges to the number of faces

      ii The number of vertices to the number of edges

    b Work out the same ratios for a triangular prism.

3 This table shows the fastest production cars that could be legally driven on public roads in different years.

Year	Model	Maximum speed (km/h)
1894	Benz Velo	19
1949	Jaguar XJ120	201
1986	Porsche 959	314
2010	Bugati Veyron 415	415

    a Work out ratios, in the form $a : 1$, to compare the speeds of the other three cars to the speed of the Benz Velo.

    b Work out a ratio to compare the speeds of the Porsche and the Bugati.

    c Work out the missing numbers to complete this sentence:

      The ratio of the speeds of the Jaguar, the Porsche and the Bugati is 1 : … : … .

    d The official world speed record for a wheel-driven car is 644.96 km/h, set by Donald Campbell in *Bluebird CN7* in 1964.

      Use a ratio to compare this to the speed of the Bugati Veyron.

4 The table shows the highest and lowest attendances for three teams in the 2012/13 football season.

Team	Highest attendance	Lowest attendance
Brighton	30 003	23 703
Middlesborough	28 229	13 377
Crewe	6547	3770

a Use ratios to compare the highest and lowest attendances for each team.

b Use a ratio with three numbers to compare the highest attendances of the three teams.

c The ratio of the largest attendance at Brighton to the largest attendance at Brentford is 2.44 : 1.

The corresponding ratio for lowest attendance is 5.41 : 1.

Work out the attendance figures for Brentford.

5 On a bicycle the large chainring on the pedals has 40 teeth and the small sprocket on the wheel has 16 teeth.

The gear ratio is the ratio of the number of teeth on the chainring to the number of teeth on the sprocket.

a Work out the gear ratio in this case. Write the answer as simply as possible.

b If the pedals go round once, how many times will the wheel rotate?

Simone's bike has three chainrings with 48, 38 and 28 teeth. There are seven sprockets on the back wheel with between 11 and 32 teeth.

c How many different gears are there on Simone's bike? (Each possible gear uses one chainring and one sprocket.)

d Work out the largest and smallest possible gear ratios for Simone's bike. Give your answers in the form $x : 1$.

6 These are the average distances of the planets in the Solar System from the Sun, in millions of kilometres.

Planet	Mercury	Venus	Earth	Mars
Distance (km)	58	108	150	228
Planet	Jupiter	Saturn	Uranus	Neptune
Distance (km)	779	1430	2880	4500

a Name two planets for which the ratio of their distances from the Sun is approximately 1 : 3.

b Name two planets for which the ratio of their distances from the Sun is approximately 2 : 3.

c Work out the ratio of the distances of the planets that are closest to and the furthest away from the Sun.

d Work out numbers to complete this sentence:

The ratio of the distances of Venus, Earth and Mars from the Sun is … : 1 : … .

Problem solving
Smoothie bar

> Recipes are for medium smoothies.

> To make a small smoothie:
> Use 60% of the ingredients in the medium recipe.

> To make a large smoothie:
> Increase the ingredients in the medium recipe by 50%

> Buy any two smoothies and get the cheaper one for half price!

Small	240 ml	£2.50
Medium	400 ml	£3.00
Large	600 ml	£4.00

1 Cost and size

   a Work out the ratio of the sizes of a medium smoothie and a large smoothie.
   b Work out the ratio of the costs of a medium smoothie and a large smoothie.
   c Work out ratios to compare the sizes and costs of a small smoothie and a large smoothie.
   d The ratios of the sizes and the ratios of the costs are not the same. Are larger sizes better value? Give a reason for your answer.

2 Ratios of ingredients

   a Work out the ratio of the masses of mango to strawberry to banana in a medium size Fruity Surprise.
   b Work out the same ratio for a large Fruity Surprise.
   c What fraction of the fruit in a Fruity Surprise is mango?
   d What fraction of the ingredients of a medium smoothie are used to make a small smoothie?
   e Work out the ratio of juice in a Fruity Surprise to juice in the same size Chocolate smoothie.

Chocolate

100 g banana
50 g chocolate spread
200 ml cranberry juice

Breakfast Boost

100 g strawberries
1 banana
150 ml milk

Tropical Fruit

250 g tropical fruit
100 ml yoghurt
85 g raspberries
Juice of a lime
1 tsp honey

Fruity Surprise

100 g mango
50 g strawberries
75 g bananas
250 ml orange juice

3 Selling smoothies

The Smoothie bar sells twice as many Breakfast Boosts as Fruity Surprises.
For every three Chocolate smoothies sold they sell four Tropical Fruit smoothies.
They sell 50% more Breakfast Boosts than Chocolate smoothies.
Work out as many different ratios as you can, to compare the sales of different types of smoothie.

4 Quantities

a One day the Smoothie bar has only 5 litres of orange juice, 5 litres of milk, 5 litres of cranberry juice and 5 litres of yoghurt. It has plenty of all the other ingredients.
Work out the largest number of each type of medium smoothie they can make.

b On another day there are enough ingredients to make 100 medium Breakfast Boosts.
  i How many large Breakfast Boosts could be made with those ingredients?
  ii How many small Breakfast Boosts could be made with those ingredients?

5 Buying smoothies

a Susie buys two smoothies. Work out how much she pays if:
  i both are large   ii one is large and one is medium
  iii one is medium and one is small.

b Karl buys 2 small, 3 medium and 4 large smoothies. How much does he pay?

c Bridget has £20.
  i Explain how she can buy 10 small smoothies with £20 and get some change.
  ii Investigate how many medium or large smoothies she can buy with £20 and what change she will get in each case.

Glossary

24-hour clock   A method of measuring the time, based on the full twenty-four hours of the day, rather than two groups of twelve hours.

3D   Having three dimensions: length, width and height.

acute angle   An angle between 0° and 90°.

add   Combine two numbers or quantities to form another number or quantity, known as the sum or total.

addition   A basic operation of arithmetic, combining two or more numbers or values to find their total value.

algebra   The use of letters to represent variables and unknown numbers and to state general rules or properties; for example: $2(x + y) = 2x + 2y$ describes a relationship that is true for any numbers x and y.

algebraic rule   A rule that connects two or more algebraic unknowns.

alternate angles   Angles that lie on either side of a transversal that cuts a pair of parallel lines; the transversal forms two pairs of alternate angles, and the angles in each pair are equal.

angle   The amount of turn between two straight lines with a common end point or vertex.

angle of rotation   The angle through which an object or shape is rotated into a new position.

angles at a point   The angles that are formed at a point where two or more lines meet; they add up to 360°.

angles on a straight line   The angles that are formed to make a straight line; they add up to 180°.

approximation   A value that is close but not exactly equal to another value, and can be used to give an idea of the size of the value; for example, a journey taking 58 minutes may be described as 'taking approximately an hour'; the sign ≈ means 'is approximately equal to'.

area   The amount of flat space a 2D shape occupies; usually measured in square units such as square centimetres (cm²) or square metres (m²).

aspect ratio   For a television screen, the ratio of the width to the height.

at random   Chosen by chance, without looking; every item has an equal chance of being chosen.

average   A central or typical value of a set of data, that can be used to represent the whole data set; mean, median, and mode are all types of average.

axis, plural axes   Fixed lines on a graph, usually perpendicular, numbered and used to identify the position of any point on the graph.

balance   The amount of money in, for example, a bank account after payments in and out have been made.

bank statement   A statement listing the amounts of money put into or taken out of a bank account.

bar chart or graph   A diagram showing quantities as vertical bars so that the quantities can be easily compared.

base   One of the sides of a 2D shape, usually the one drawn first, or shown at the bottom of the shape.

biased   Not random, for example, attaching a small piece of sticky gum to the edge of a coin may cause it to land more frequently on Heads than on Tails: the coin would be biased.

BIDMAS   An agreed order of carrying out operations Brackets, Indices (or powers), Division and Multiplication, Addition and Subtraction.

brackets   Symbols used to show expressions that must be treated as one term or number. Under the rules of BIDMAS, operations within brackets must be done first; for example: $2 \times (3 + 5) = 2 \times 8 = 16$ whereas $2 \times 3 + 5 = 6 + 5 = 11$.

calculate   Work out, with or without a calculator.

capacity   The volume of a liquid or gas inside a 3D shape, usually measured in litres.

cent- or centi-   A prefix referring to 100; in the metric system, centi- means one hundredth; for example, a centimetre is one hundredth of a metre.

centre of rotation   The point about which an object or shape is rotated.

chance   The likelihood, or probability, of an event occurring.

chart   A diagram or table showing information.

Chinese method   A method of performing multiplication by arranging the numbers in a particular layout.

class   A small range of values within a large set of data, treated as one group of values.

coefficient   A number written in front of a variable in an algebraic term; for example, in $8x$, 8 is the coefficient of x.

column method   A method for multiplying large numbers, in which you multiply the units, tens and hundreds separately, then add the products together.

combined events   Two or more events (independent or mutually exclusive) that may occur during a trial.

common factor   A factor that divides exactly into two or more numbers; 2 is a common factor of 6, 8 and 10.

compound shape   A shape made from two or more simpler shapes; for example, a floor plan could be made from a square and a rectangle joined together.

construct   Draw angles, lines or shapes accurately, using compasses, a protractor and a ruler.

continuous data   Data that can take any value, such as height or mass.

conversion   Expression of a unit or measurement in terms of another unit or scale of measurement.

conversion graph   A graph that can be used to convert from one unit to another, constructed by drawing a line through two or more points where the equivalence is known; sometimes, but not always, a conversion graph passes through the origin.

convert   **1** Express a unit or measurement in terms of another unit or scale of measurement; for example, you can convert inches to centimetres by multiplying by 2.54.
2 To change a number from one form to another, for example, from a fraction to a decimal.

coordinates   Pairs of numbers that show the exact position of a point on a graph, by giving the distance of the point from each axis; on an x–y coordinate graph, in the set of coordinates (3, 4), 3 is the x-coordinate, and is the horizontal distance of the point from the y-axis, and 4 is the y-coordinate, and is the vertical distance of the point from the x-axis.

corresponding angles   Angles that lie on the same side of a pair of parallel lines cut by a transversal; the transversal forms four pairs of corresponding angles, and the angles in each pair are equal.

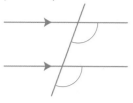

cube **1** In geometry, a 3D shape with six square faces, eight vertices and 12 edges.
2 In number and algebra, the result of multiplying a number or expression raised to the power of three: n^3 is read as 'n cubed' or 'n to the power of three': for example: 2^3 is the cube of 2 and $(2 \times 2 \times 2) = 8$.

cuboid A 3D shape with six rectangular faces, eight vertices and 12 edges; opposite faces are identical to each other.

data A collection of facts, such as numbers, measurements, opinions or other information.

data-collection form A form or table used for recording data collected during a survey.

decimal A number or number system that is based on 10; a decimal number usually means a number made up of a whole number and fractions, expressed as tenths, hundreds, thousandths, written after a decimal point.

decimal place The position, after the decimal point, of a digit in a decimal number; for example, in 0.025, 5 is in the third decimal place. Also, the number of digits to the right of the decimal point in a decimal number; for example, 3.142 is a number given to three decimal places (3 dp).

decimal point A symbol, usually a small dot, written between the whole-number part and the fractional part in a decimal number.

decrease **1** Reduce or make smaller.
2 The amount by which something is made smaller.

degree A measure of angle equal to $\frac{1}{360}$ of a complete turn.

denominator The number below the line in a fraction, which says how many parts there are in the whole; for example, a denominator of 3 tells you that you are dealing with thirds.

deposit Money paid as a first instalment on the purchase of something; the balance must be paid later.

diagonal A straight line joining any two non-adjacent vertices of a polygon.

direction of rotation The direction in which a shape is rotated, either clockwise or anticlockwise.

discrete data Data that can take only certain values, such as numbers of children in a family.

dodecahedron A 3D shape with 12 faces.

edge The line where two faces or surfaces of a 3D shape meet.

equally likely When the probabilities of two or more outcomes are equal; for example, when a fair six-sided die is thrown, the outcomes 6 and 2 are equally likely with probabilities of $\frac{1}{6}$.

equation A number sentence stating that two expressions or quantities are of equal value, for example, $x + 2y = 9$; an equation always contains an equals sign (=).

equivalent The same, equal in value.

equivalent fraction Fractions that can be cancelled to the same value, such as $\frac{10}{20} = \frac{5}{10} = \frac{1}{2}$.

estimate **1** State or guess a value, based on experience or what you already know.
2 A rough or approximate answer.

Euler Leonhard Euler (1707–83), a Swiss mathematician.

Euler's formula A formula connecting the numbers of faces (F), vertices (V) and edges (E) of a 3D shape: $F + V - E = 2$.

event Something that happens, such as the toss of a coin, the throw of a dice or a football match.

experiment A test or investigation made to find evidence for or against a hypothesis.

experimental probability The probability found by trial or experiment; an estimate of the true probability.

expression Collection of numbers, letters, symbols and operators representing a number or amount; for example, $x^2 - 3x + 4$.

exterior angle An angle formed by extending one side of a 2D shape at a vertex, to create an angle outside the shape.

Fibonacci sequence A sequence of numbers in which the third and subsequent terms are formed by adding the two previous terms: 1, 1, 2, 3, 5, 8, …

face One of the flat surfaces of a solid shape; for example, a cube has six faces.

fair The probability of each outcome is similar to the theoretical probability.

first term The first number in a sequence.

formula A mathematical rule, using numbers and letters, that shows how to work out an unknown variable; for example, the conversion formula from temperatures in Fahrenheit to temperatures in Celsius is: $C = \frac{5}{9}(F - 32)$.

formulae The plural form of formula.

fraction A part of a whole that has been divided into equal parts, a fraction describes how many parts you are talking about.

fraction wall A diagram that allows you to compare fractions and see which ones are equivalent.

frequency The number of times a particular item appears in a set of data.

frequency table A table showing data values, or ranges of data values, and the numbers of times that they occur in a survey or trial.

function A mathematical operation that changes a set of numbers into another set of numbers.

function machine A diagram to illustrate functions and their inputs and outputs.

geometric sequence A sequence in which each term is multiplied or divided by the same number, to work out the next term; for example, 2, 4, 8, 16, ... is a geometric sequence.

geometrical properties The properties of a 2D or 3D shape that describe it completely.

graph A diagram showing the relation between certain sets of numbers or quantities by means of a series of values or points plotted on a set of axes.

greater than, > The symbol > shows that the amount on the left of it is greater or more than the amount on the right of it.

grid or box method A method for multiplying numbers larger than 10, where each number is split into its parts: for example, to calculate 158×67:

   158 is 100, 50 and 8

   67 is 60 and 7.

These numbers are arranged in a rectangle and each part is multiplied by the others.

grouped data   Data that is arranged into smaller, non-overlapping sets, groups or classes, that can be treated as separate ranges or values, for example, 1–10, 11–20, 21–30, 31–40, 41–50; in this example there are equal class intervals.

grouped frequency table   A table showing data grouped into classes.

height   The vertical distance, from bottom to top, of a 2D or 3D shape.

hexagonal prism   A prism with a hexagonal cross-section and six rectangular faces; it has 8 faces, 12 vertices and 18 edges.

hexomino   A 2D shape that is made from six identical squares.

hypothesis   A theory or idea that may be tested.

icon   A symbol or graphic representation on a chart or graph.

image   The result of a reflection or other transformation of an object.

improper fraction   A fraction in which the numerator is greater than the denominator. The fraction could be rewritten as a mixed number for example, $\frac{7}{2} = 3\frac{1}{2}$.

increase   **1** Enlarge or make bigger.
2 The amount by which something is made bigger.

input   The number put into a function machine.

interest   Money earned by someone who deposits or lends a sum of money to someone else; you can earn interest by putting money into a bank savings account, but you will be charged interest if you borrow money from a bank or building society.

interior angle   An angle formed inside a 2D shape, where two sides meet at a vertex.

intersect   To have a common point for example, two non-parallel lines cross or intersect at a point.

inverse   Reverse or opposite; inverse operations cancel each other out or reverse the effect of each other.

inverse operation   An operation that reverses the effect of another operation; for example, addition is the inverse of subtraction, division is the inverse of multiplication.

isometric   A grid of equilateral triangles or dots, can be used for drawing a 3D shape in 2D.

isosceles triangle   A triangle in which two sides are equal and the angles opposite the equal sides are also equal.

length   The distance from one end of a line to the other.

less than, <   The symbol < shows that the amount on the left of it is smaller or less than the amount on the right of it.

like terms   Terms in which the variables are identical, but have different coefficients; for example, $2ax$ and $5ax$ are like terms but $5xy$ and $7y$ are not. Like terms can be combined by adding their numerical coefficients so $2ax + 5ax = 7ax$.

line graph   A chart that shows how data changes, by means of points joined by straight lines.

line of symmetry   A line that divides a symmetrical shape into two identical parts, one being the mirror image of the other.

linear sequence   A sequence or pattern of numbers where the difference between consecutive terms is always the same.

list of outcomes   A list of all possible results (outcomes) of an event or trial.

litre   A metric measure of capacity; 1 litre = 1000 millilitres = 1000 cubic centimetres.

long division   A method of division showing all the workings, used when dividing large numbers.

long multiplication   A method of multiplication showing all the workings, used when multiplying large numbers.

lowest common multiple (LCM)   The lowest number that is a multiple of two or more numbers; 12 is the lowest common multiple of 2, 3, 4 and 6.

mean   An average value of a set of data, found by adding all the values and dividing the sum by the number of values in the set; for example, the mean of 5, 6, 14, 15 and 45 is $(5 + 6 + 14 + 15 + 45) \div 5 = 17$.

mean average   *See* mean.

median   The middle value of a set of data that is arranged in order; for example, write the data set 4, 2, 6, 2, 2, 3, 7 in order as 2, 2, 2, 3, 4, 6, 7, then the median is the middle value, which is 3. If there is an even number of values the median is the mean of the two middle values; for example, 2, 3, 6, 8, 8, 9 has a median of 7.

metric   A system of measurement in which the basic units of mass, length and capacity are grams, metres and litres. Sub-units are obtained from main units by multiplying or dividing by 10, 100, 1000, …. For example, for mass, 1 kilogram = 1000 grams; for length, 1 kilometre = 1000 metres, 1 metre = 100 centimetres, 1 centimetre = 10 millimetres; for capacity, 1 litre = 1000 millilitres.

metric units   Units of measurement used in the metric system; for example, metres and centimetres (length), grams and kilograms (mass), litres (capacity).

mill-   Thousand, as in millennium, a thousand years.

milli-   A prefix used in the metric system of measurement to indicate a thousandth part, for example, a millimetre is one thousandth of a metre.

mirror line   Another name for a line of symmetry.

mixed number   A number written as a whole number and a fraction; for example, the mixed number $2\frac{1}{2}$ can be written as the improper fraction $\frac{5}{2}$.

modal   The value that occurs most frequently in a given set of data.

modal class   In grouped data, the class with the highest frequency.

mode   The value that occurs most frequently in a given set of data.

negative number   A number that is less than zero.

net   A 2D shape that can be folded up to make a 3D shape.

nth term   An expression in terms of n; it allows you to find any term in a sequence, without having to use a term-to-term rule.

numerator   The number above the line in a fraction: it tells you how many of the equal parts of the whole you have; for example, $\frac{3}{5}$ of a whole is made up of three of the five equal parts. The number of equal parts is the denominator.

object   The original or starting shape, line or point before it is transformed to give an image.

obtuse angle   An angle that is greater than 90° but less than 180°.

octahedron   A 3D shape with eight faces.

operation   An action carried out on or between one or more numbers; it could be addition, subtraction, multiplication, division or squaring.

opposite angles   The angles on the opposite side of the point of intersection when two straight lines cross, forming four angles. The opposite angles are equal.

order Arrange numbers or quantities according to a rule, such as size or value.

order of operations The order in which mathematical operations should be done.

order of rotational symmetry The number of times a 2D shape looks the same as it did originally when it is rotated through 360°. If a shape has no rotational symmetry, its order of rotational symmetry is 1, because every shape looks the same at the end of a 360° rotation as it did originally.

origin The point (0, 0) on Cartesian coordinate axes.

outcome The result of an event or trial in a probability experiment, such as the score from a throw of a dice.

outlier In a data set, a value that is widely separated from the main cluster of values.

output The number produced by a function machine.

parallel Lines that are always the same distance apart, however far they are extended.

parallelogram A quadrilateral with two pairs of parallel sides; the opposite sides are equal in length.

Pascal's triangle A triangular array of numbers, starting with 1 at the top; every subsequent row has an extra term, and the numbers are placed to align with the spaces in the row above. Every term is the sum of the two terms diagonally above it on either side.

```
            1
         1     1
       1    2    1
     1    3    3    1
   1    4    6    4    1
 1    5   10   10    5    1
```

pentagonal prism A prism with a pentagonal cross-section and five rectangular faces; it has 7 faces, 10 vertices and 15 edges.

pentomino A shape made by joining five squares together side-to-side. There are 12 such shapes.

per cent (%) Parts per hundred.

percentage A number written as a fraction with 100 parts, but instead of writing it as a fraction out of 100, you write the symbol % at the end, so $\frac{50}{100}$ is written as 50%.

perimeter The total distance around a 2D shape; the perimeter of a circle is called the circumference.

perpendicular At right angles, meeting at 90°.

perpendicular height The distance between the base of a 2D shape and its topmost point or vertex, measured at right angles to the base.

pictogram A method of displaying data, using small pictures or icons to represent one, two or more items of data.

pie chart A circular graph divided into sectors that are proportional to the size of the quantities represented.

place value The value of a digit depending on where it is written in a number; for example, in the number 123.4, the place value of 4 is tenths, so it is worth 0.4 and the place value of 2 is tens, so it is worth 20.

positive number A number that is greater than zero.

power How many times you use a number or expression in a calculation; it is written as a small, raised number; for example, 2^2 is 2 multiplied by itself, $2^2 = 2 \times 2$ and 4^3 is $4 \times 4 \times 4$.

probability The measure of how likely an outcome of an event is to occur. All probabilities have values in the range from 0 to 1.

probability fraction A probability that is not 0 or 1, given as a fraction.

probability scale A scale or number line, from 0 to 1, sometimes labelled with impossible, unlikely, even chance,

etc., to show the likelihood of an outcome of an event occurring. Possible outcomes may be marked along the scale as fractions or decimals.

protractor A transparent circular or semicircular instrument for measuring or drawing angles, graduated in degrees.

quadrant One of the four regions into which a plane is divided by the coordinate axes in the Cartesian system.

quadrilateral A 2D shape with four straight sides. Squares, rhombuses, rectangles, parallelograms, kites and trapezia are all special kinds of quadrilaterals.

quantity A measurable amount of something that can be written as a number, or a number with appropriate units; for example, the capacity of a milk carton.

questionnaire A list of questions for people to answer, so that statistical information can be collected.

random Chosen by chance, without looking; every item has an equal chance of being chosen.

range The difference between the greatest value and the smallest value in a set of numerical data. A measure of spread in statistics.

ratio A way of comparing the sizes of two or more numbers or quantities; for example, if there are five boys and ten girls in a group, the ratio of boys to girls is 5 : 10 or 1 : 2, the ratio of girls to boys is 2 : 1. The two numbers are separated by a colon (:).

reciprocal The result of dividing a number into 1; the reciprocal of N is $\frac{1}{N}$.

rectangle A quadrilateral in which all four interior angles are 90° and two pairs of opposite sides are equal and parallel; it has two lines of symmetry and rotational symmetry of order 2. The diagonals of a rectangle bisect each other.

reduce Make less, or make smaller.

reduction *See* decrease.

reflect Draw an image of a 2D shape as if it is viewed in a mirror placed along a given (mirror) line.

reflection The image formed when a 2D shape is reflected in a mirror line or line of symmetry; the process of reflecting an object.

reflective symmetry A type of symmetry in which a 2D shape is divided into two equal parts by a mirror line.

reflex angle An angle that is greater than 180° but less than 360°.

relationship An association between two or more items.

remainder The amount left over after dividing a number.

repeated subtraction A type of division involving the process of repeatedly subtracting the same number or amount; for example, $35 - 5 - 5 - 5 - 5 - 5 - 5 - 5 = 0$ so $35 \div 5 = 7$, remainder 0.

right angle One quarter of a complete turn. An angle of 90°.

rotate Turn.

rotation How a 2D shape is rotated.

rotational symmetry A type of symmetry in which a 2D shape may be turned through 360° so that it looks the same as it did originally in two or more positions.

round In the context of a number, to express to a required degree of accuracy; for example, 653 rounded to the nearest 10 is 650.

round down To change a number to a lower and more convenient value; for example, 451 rounded down to the nearest ten is 450.

round up To change a number to a higher and more convenient value; for example, 459 rounded up to the nearest ten is 460.

rounding Expressing to a required degree of accuracy; for example, 743 rounded to the nearest 10 is 740.

rule   The way a mathematical function is carried out. In patterns and sequences a rule, expressed in words or algebraically, shows how the pattern or sequence grows or develops.

sample   A selection taken from a larger data set, which can be researched to provide information about the whole data set.

sample space   The set of all possible outcomes for an event or trial.

scaling   A method used in drawing statistical diagrams, such as pie charts; data values are multiplied or divided by the same number, so that they can be represented proportionally in a diagram.

sector   A region of a circle, like a slice of a pie, bounded by an arc and two radii.

sequence   A pattern of numbers that are related by a rule.

short division   The division of one number by another, usually an integer, that can be worked out mentally rather than on paper.

significant figure   In the number 12 068, 1 is the first and most significant figure and 8 is the fifth and least significant figure. In 0.246 the first and most significant figure is 2. Zeros at the beginning of a number are not significant figures.

simplest form   **1** A fraction that has been cancelled as much as possible.
2 An algebraic expression in which like terms have been collected, so that it cannot be simplified any further.

simplify   To make an equation or expression easier to work with or understand by combining like terms or cancelling; for example, $4a - 2a + 5b + 2b = 2a + 7b$, $\frac{12}{18} = \frac{2}{3}$, $5 : 10 = 1 : 2$.

solve   To find the value or values of a variable (x) that satisfy the given equation.

square   A quadrilateral in which all four interior angles are 90° and all four sides are equal; opposite sides are parallel, the diagonals bisect each at right angles; it has four lines of symmetry and rotational symmetry of order 4.

square number   A number that results from multiplying an integer by itself; for example, $36 = 6 \times 6$ and so 36 is a square number. A square number can be represented as a square array of dots.

square root   For a given number, a, the square root is the number b, where $a = b^2$; for example, a square root of 25 is 5 since $5^2 = 25$. The square root of 25 is recorded as $\sqrt{25} = 5$. Note that a positive number has a negative square root, as well as a positive square root; for example, $(-5)^2 = 25$ so it is also true that $\sqrt{25} = -5$.

squaring   Multiplying a number or expression by itself; raising a number or expression to the second power; for example, $3^2 = 9$.

statistical survey   The collection of statistical information.

straight-line graph   A graph of a linear function or equation, such as $y = 2x + 3$, for which all the points lie in a straight line.

substitute   Replace a variable in an expression with a number and evaluate it; For example, if we substitute 4 for t in $3t + 5$ the answer is 17 because $3 \times 4 + 5 = 17$.

subtraction   Taking one number or quantity away from another, to find the difference.

surface area   The total area of all of the surfaces of a 3D shape.

tally   A mark made to record a data value; every fifth tally is drawn through the previous four.

tally chart   A chart with marks made to record each object or event in a certain category or class. The marks are usually grouped in fives to make counting the total easier.

term   **1** A part of an expression, equation or formula. Terms are separated by + and – signs.
2 A number in a sequence or pattern.

term-to-term rule   The rule that shows what to do to one term in a sequence, to work out the next term.

tessellation   A pattern made of one or more repeating shapes that fit together without leaving any gaps between them.

tetrahedron   A 3D shape with four triangular faces; in a regular tetrahedron, the faces are equilateral triangles. A tetrahedron has 4 faces, 4 verticals and 6 edges.

theoretical probability   The probability of an outcome of an event or trial, based on calculation.

timetable   A table showing when events take place.

trailing zeros   Zeros written at the end of a decimal fraction, to act as place-holders, but which do not change the value of the decimal number.

transversal   A straight line that cuts two or more parallel lines.

trapezium   A quadrilateral in which only one pair of opposite sides are parallel but unequal in length. In an isosceles trapezium, the other two sides are the same length as each other.

trial   An experiment to discover an approximation for the probability of an outcome of an event; it will consist of many trials where the event takes place and the outcome is recorded.

triangle   A 2D shape with three straight sides; the interior angles add up to 180°. Triangles may be classified as:
- scalene – no sides are equal, no angles are equal
- isosceles – two of the sides are equal, two of the angles are equal
- equilateral – all the sides are equal, all the angles are equal
- right-angled – one interior angle is equal to 90°.

triangular number   A number in the sequence 1, 1 + 2, 1 + 2 + 3, 1 + 2 + 3 + 4, …. 55 is a triangular number since $55 = 1 + 2 + 3 + 4 + 5 + 6 + 7 + 8 + 9 + 10$. A triangular number can be represented by a triangular array of dots, in which the number of dots increases by 1 in each row.

triangular prism   A prism with a triangular cross-section and three rectangular faces; it has 5 faces, 6 vertices and 9 edges.

units digit   The digit that appears furthest right in a whole number, or before the decimal point in a decimal number; for example, in 315, the units digit is 5 and in 123.4 the units digit is 3.

unknown number   A number that is represented by a letter; it can be treated as a number, following the rules of arithmetic (BIDMAS).

unlike terms   Terms that are not made up from the same variables, such as 5, x, y, xy. You cannot combine unlike terms.

variable   A quantity that may take many values.

vertex   The point at which two lines meet, in a 2D or 3D shape.

vertically opposite angles   *See* opposite angles.

vertices   The plural of vertex.

volume   The amount of space occupied by a 2D shape.

width   The distance from one side of a 2D shape to the other, usually taken to be shorter than the length.

x-axis   The horizontal axis of a two-dimensional x–y Cartesian coordinate graph, along which the x-coordinates are measured.

x-coordinate   The horizontal distance of the point from the y-axis; the position of a point along the x-axis.

y-axis   The vertical axis of a two-dimensional x–y Cartesian coordinate graph, along which the y-coordinates are measured.

y-coordinate   The vertical distance of the point from the x-axis; the position of a point up the y-axis.

Index